More praise for
EMBLEMS OF MIND

"Very rewarding . . . a profound book that has much to say about both music and mathematics."

—*Los Angeles Times*

"A powerful and provocative work . . . Regular readers of Edward Rothstein's criticism will not be surprised by the thoughtfulness of his approach to the problem of beauty in music . . . Few books have taught me more about the essential nature of music, and fewer still have done so with comparable grace and clarity."

—Terry Teachout, *Commentary*

"Passionate . . . fascinating . . . Rothstein's explanations are clear and he is remarkably skilled in drawing readers into the fold . . . **EMBLEMS OF MIND** is his personal journey. It is a quest that has given him enormous pleasure, and he welcomes all those who wish to join him on the road."

—*Washington Post Book World*

"Mr. Rothstein's splendid and lucid book reveals a narrative prose so engaging and a thesis so inciting to reflection that it should delight anybody interested in either (or both) music and mathematics."

—Jacques Barzun, author of *The Pleasures of Music*

"A fine new book . . . Like a great piece of music, it is initially rewarding and yet invites a return visit."

—*Strings*

EMBLEMS OF MIND

The
INNER LIFE OF MUSIC
AND MATHEMATICS

EDWARD ROTHSTEIN

AVON BOOKS  NEW YORK

Grateful acknowledgment is made to the following for permission to reprint previously published material. BIRKHÄUSER: Quotations and examples from *The Mathematical Experience* by Philip J. Davis and Reuben Hersh. Reprinted by permission of Birkhäuser, Cambridge, MA. DOVER BOOKS: Illustrations from *Treatise on Harmony* by Rameau, translated by Philip Gossett; diagram of Bach's Prelude No. 1 in C by Heinrich Schenker in *Five Graphic Music Analyses*; illustrations from *The Geometry of Art and Life* by Matila Ghyka. Reprinted by permission. EUROPEAN AMERICAN MUSIC: Webern II. *Kantate*, Op. 31. Copyright © 1956 (renewed) by Universal Edition A. G., Wien. All rights reserved. Reprinted by permission of European American Music Distributors Corporation, sole U.S. and Canadian agent for Universal Edition A. G., Wien. HARPERCOLLINS PUBLISHERS, INC.: Diagram based on illustration from page 464 from *The Republic of Plato* translated by Allan Bloom. Reprinted by permission of Basic Books, a division of HarperCollins Publishers, Inc. THE NEW YORK TIMES: "Math and Music: The Deeper Links" by Edward Rothstein (August 29, 1982). Copyright © 1982 by The New York Times. Reprinted by permission. SPRINGER-VERLAG, INC.: Excerpt from an article by Rozsa Peter in *The Mathematical Intelligencer*, vol. 12, no. 1. Reprinted by permission of Springer-Verlag New York, Inc.

AVON BOOKS
A division of
The Hearst Corporation
1350 Avenue of the Americas
New York, New York 10019

Copyright © 1995 by Edward Rothstein
Published by arrangement with Times Books, a division of Random House, Inc.
Library of Congress Catalog Card Number: 94-171
ISBN: 0-380-72747-1

The Times Books edition contains the following Library of Congress Cataloging in Publication Data:
Rothstein, Edward.
    Emblems of mind: the inner life of music and mathematics / Edward Rothstein.
        p.   cm.
Includes index.
1. Music—Philosophy and aesthetics.   2. Music—Theory—Mathematics.
3. Mathematics—Philosophy.   I. Title.
ML3800.R62   1995
780´.051—dc20   94-171

First Avon Books Trade Printing: August 1996

AVON TRADEMARK REG. U.S. PAT. OFF. AND IN OTHER COUNTRIES, MARCA REGISTRADA, HECHO IN U.S.A.

Printed in the U.S.A.

OPM   10   9   8   7   6   5   4   3   2

To Dena, who grew up with it.
To Aaron, who was born with it.
To Anna, who came at its conclusion.
And to Marilyn, who waited.

All Nature is but Art, unknown to thee;
All Chance, Direction, which thou canst not see;
All Discord, Harmony not understood ...

—ALEXANDER POPE

ACKNOWLEDGMENTS

FOR A BOOK SO CONCERNED WITH ABSTRACTION AND BEAUTY, this project has incurred a great many worldly debts along with intellectual and aesthetic ones. Given its subject, I must begin by thanking my teachers who over the years have taught me how to think, whether at the keyboard (Maria Cisyk, Patricia Zander, and Lisa Grad) or at the blackboard (Mordechai Friedman, Shizuo Kakutani, Abraham Robinson). My experience at the Committee on Social Thought at the University of Chicago disclosed so many different styles of intelligence and encouraged such devotion to great texts and their meanings that the writing of this book was inseparable from the experience of being at "the Committee" where I combined studies of music, philosophy, and literature. My teachers there—Saul Bellow, Edward Shils, Stephen Toulmin, Paul Ricoeur, David Grene, Allan Bloom, Leon Kass, James Redfield—along with their predecessors in other schools, influenced many passages of this book.

This project, despite its abstract cast, actually grew out of the world of daily journalism where during my first stint as a music critic of *The New York Times*, Arthur Gelb, then managing editor of the paper, suggested I write an essay about my two strongest intellectual passions. Every conversation with Arthur led to something unexpected; this time an essay on music and mathematics in the Arts & Leisure section of the Sunday *Times* was the result. This book is an attempt to grasp with more completeness the suggestions I made in that essay.

I could never have completed the work without the generous support and encouragement of Marvin and Margaret Menzin, who believed in the project and who remained constant in their conviction. Margaret, a mathematician at Simmons College, was an astute and careful commentator on the manuscript as well.

Daniel and Joanna S. Rose also provided much-needed assistance, which was all the more remarkable for having been unsought and given without condition. Their grant was administered by the New York Foundation for the Arts.

The Ingram Merrill Foundation also provided a valued grant for part of the period during which this book was written.

In addition, during my years of freelance work in the writing market, juggling magazine assignments, columns, reporting, feature profiles, and this project, the New York Institute for the Humanities provided a physical and intellectual home. A. Richard Turner was the institute's director, Jocelyn Carlson provided warm friendship and selfless assistance, and my fellow fellows provided many occasions for argument, companionship, and relief from the blinking computer cursor.

My agent, Robert Cornfield, was interested in this project from the moment he heard about it and then remained committed to it, guiding it through two separate journeys through the corridors of publishing houses. It finally ended up in the good hands of Steve Wasserman, editorial director of Times Books, who has been an ideal editor: enthusiastic, imaginative, firm, and intelligent. He improved this book markedly with his suggestions and insights.

At various times, the manuscript was read by many people, including Neal Kozodoy, whose friendship did not cloud his objective scrutiny; John Hollander, who expressed early and avid interest; and Milton Babbitt, whose intelligence, enthusiasms, and critical faculties are as evident in his conversations as in his compositions. There were valuable suggestions as well from Jacques Barzun, Burton Fine, Martin Gardner, Joseph Kerman, Louis Menand, and Sidney Morgenbesser. Stephen Toulmin and Charles Rosen read the manuscript as members of the faculty of the Committee on Social Thought; they were invaluable advocates and critics. David Lewin gave the manuscript two indispensable close readings, refining its prose and points. I have tried to incorporate the suggestions of these many distinguished readers. The errors that remain, of course, are mine alone.

Finally, only my wife, Marilyn, knows the true cost of devotion to such issues as this book obsesses over; even tentative gropings after beauty and truth come at a large price, which she often had to bear alone. My parents, Joseph and Phyllis Rothstein, have had long experience in such matters; it is under their guidance that I first became interested in these fields. And my children—Dena and Aaron and now Anna—will have to manage without the constant shadow of this project, which has been around almost as long as their love. To Marilyn and to my children, my deep thanks for the patience they displayed waiting for this book and for their unquestioning acceptance of its flawed, all-too-human reality.

CONTENTS

INTRODUCTION

B EFORE SETTING OUT TO MAKE MY WAY IN THE MUSIC BUSINESS, I was in training to become a "pure" mathematician. Such esoteric subjects as algebraic topology, measure theory, and nonstandard analysis were my preoccupations. I would stay up nights trying to solve knotty mathematical problems, playing with abstract phrases and structures. But at the same time, I would be lured away from these constructions by another activity. With an enthusiasm that could come only when critical faculties are in happy slumber, I would listen to or play a musical composition again and again, imprinting my ear and mind and hands with its logic and sense. Music and math together satisfied a sort of abstract "appetite," a desire that was partly intellectual, partly aesthetic, partly emotional, partly, even, physical.

These facts are not extraordinary among those who have been involved with these fields. Not only did I know other people tempted by both worlds, but, in various ways, music and mathematics have been associated throughout history. Mathematicians and physicists of all epochs have felt the affinities. Galileo speculated on numerical reasons "why some combinations of tones are more pleasing than others." Euclid wondered about those combinations some two thousand years earlier. The eighteenth-century mathematician Leonhard Euler wrote a discourse on the relationship of consonance to whole numbers (and one of Euler's books inspired a contemporary comment that it "contained too much geometry for musicians and too much music for geometers"). Johannes Kepler believed the planets' revolutions literally created a "music of the spheres"—a sonic counterpart to his mathematical laws of planetary motion.

Musicians, on the other hand, have invoked mathematics to describe the orderliness of their art. Chopin said, "The fugue is like pure logic in music." Bach, the fugue's most eminent explorer, also had a predilection for its precise relative, the canon, which he often treated as a puzzle. In the twentieth century, mathematical language has pervaded much musical thinking. Schoenberg's serial system for manipulating the scale's twelve tones has exercised enormous influence. Musicologists have invoked "set theory," "Markov chains," and other mathematical concepts. Journal articles detail attempts to decompose, perform, and compose music using computer programs. Iannis Xenakis applies sophisticated mathematical theories in his compositions. Even John Cage, in his search for lack of order, used computer-generated random numbers for composing.

This book is an attempt to explain why these connections are far from accidental or incidental and why they reveal something profound about the nature of each activity. The method I have chosen is, admittedly, peculiar. I do not provide a lot of anecdotes or write breezily about coincidental similarities. This is not a historical survey, dutifully beginning with Greece and extending through the Renaissance to the present. Nor do I make an attempt to be systematic in my description of these activities; this is not an academic introduction to these subjects. I do not even pretend to have solved the problems I have posed. This book is theoretical but it does not present a theory; it is analytical without being about analysis; it is mathematical but never probes too deeply into mathematics. In fact, probably very little of what I say about mathematics will be news to mathematicians, and very little of what I say about music will be news to musicians and composers. The hope is that much of what I have to say will still be of interest because of the juxtapositions I make and the hypotheses I propose.

What I have tried to do is to give some sense of what it is actually like to be immersed in both activities, what qualities of mind they demand, and how similar they are in the midst of their differences. I do not mean to press similarities into identities, or to minimize the often essential distinctions that must be maintained between a science and an art. The chapters that follow are speculations, considerations, part of an intellectual essay that is an essay in the sense of the French origins of the word: as an "attempt."

My intention has been to write for the literate layman, the curious listener and student who might never have studied math beyond high school or music beyond elementary school. I have chosen examples in mathematics that can be understood fairly easily and examples in music that are so familiar there might be no need to hear the scores themselves. There are times when the exposition might seem too demanding; in that case, I hope the reader might simply pass over the material. I have tried to write the book so that no particular example is essential. The point is less in the details than in the argument's direction.

This book is also slightly old-fashioned, for I try, in its final chapters, to discuss such unfashionable concepts as beauty and truth. I do not pretend to be defining timeless standards for either, but in the process of giving our due to history and culture, we have slighted the powers of the timeless, the necessary, and the absolute. My hope is that by the end of this book the nature of the "high" in that most contested of terms, "High Art," will also be a bit more clear.

I have chosen a form of exposition that may at first seem

puzzling. It moves back and forth between mathematics and music, attempting to take each on their own terms while making connections between them. Some of these associations are slight (the chapter titles, for example, are metaphorical allusions to the character of musical compositions). Some, though, are meant to penetrate more deeply. Part of what I want to show in this book is how exploration proceeds in music and mathematics; I attempt to evoke a feeling for the internal life of the two subjects, their reliance on metaphor, abstraction, and comparison.

The first chapter, "Prelude," is an introduction. I take as a guiding metaphor for this intellectual exploration the journey described by the poet William Wordsworth in *The Prelude*. His difficult climb to the top of Mt. Snowdon ends with a vision of the interrelationships between objects in the natural world and the mind of the observer—something which I hope this book will end with as well. I also mean this inquiry to have a "poetic" aspect; it requires constant invocation of metaphor and analogy to be effective. Why, I then ask, should mathematics and music be such constant companions through history? When we look closely, we find many obvious links, including a systematic logic that guides musical systems.

The second chapter, "Partita," attempts to explore mathematics on its own terms, to take a particular type of mathematical problem and follow its evolution through history, trying to tease out from the spirited debates and explorations some of the procedures of mathematics. I try to show how mathematics develops out of worldly questions and proceeds into ever more artful and abstract explorations. The point is not the particulars of the subject—the nature of smoothness and continuity in space and in the world of number—but the kind of approach mathematics demands, the way it keeps stepping back from the arguments it has made, discovering ever more elaborate abstractions that reveal ever deeper similarities.

The third chapter, "Sonata," attempts to do for music what

the second chapter did for mathematics: take it on its own terms and try to discern the patterns of thought that take place in musical composition. My ambition is quite general, but I try again to come at the results through specific examples, asking how music is organized and perceived. I suggest that there are many crucial similarities between the kinds of thinking that take place in understanding music and doing mathematics.

The exploration then steps back from the kinds of systems mathematics and music set up. In the fourth chapter, "Theme and Variations," the theme is, quite simply, beauty. We associate beauty with music but not often enough with mathematics. I try to analyze what its character is in each activity, and by invoking Kant, sketch some tentative universal characteristics that may govern our judgment of beauty.

The fifth chapter, "Fugue," attempts to give the same treatment to the issue of truth. I attempt to examine in an introductory fashion, how such abstract activities as music and mathematics achieve the power and meanings they do, how they can seem so otherworldly yet have such extraordinary impact on our daily lives.

Finally, the sixth chapter, "Chorale," is meant to have the same relationship to the whole as the chorale does in a Bach cantata: a statement of overarching belief and principles. It is an allusive attempt to describe the kind of argument that takes place in mathematics, music, and in this book as well. I offer no final answers, only an evocation of the kind of understanding that is a shared goal of both activities.

Stravinsky said that the musician should find in mathematics a study "as useful to him as the learning of another language is to a poet." In discussing "the art of combination which is composition" the composer quoted the mathematician Marston Morse: "Mathematics are the result of mysterious powers which no one understands, and in which the unconscious recognition of beauty must play an important part. Out of an infinity of designs a mathematician chooses one pattern for beauty's sake and pulls it down to earth." Morse, Stravinsky says, could as well have been talking about music. It is not only in the clarity of things but in their beauty and mystery that the two arts join. It is where I hope in the course of this journey we will join them as well.

EMBLEMS
OF MIND

I

PRELUDE:
THE NEED FOR METAPHOR

Ten . . . This number was of old held high in honor,
for such is the number of fingers by which we count.

OVID

INTENT ON SEEING THE SUN RISE FROM THE TOP OF MT. SNOWDON, the young William Wordsworth set out on a climb one evening two centuries ago with a friend and a shepherd guide. It was a close, warm summer night, the fog hanging low, air dripping with moisture. Beginning from a cottage at the mountain's base, the trio climbed in silence as the mists surrounded them. The poet's head was bent earthward, as if, he writes, it were set against an enemy. He was lost in thought, negotiating rocks and paths, panting breathlessly, leading the way through the midnight hours. Gradually, though dawn had not yet come, the ground at the poet's feet began to brighten. With each step the light increased. There was hardly time to ask or learn the cause, when suddenly—"Lo!" the poet cries in biblical fashion—he looked up and there was the moon

> *hung naked in a firmament*
> *Of azure without cloud, and at my feet*
> *Rested a silent sea of hoary mist.*

From the mountain peak, the poet saw a vast sea of vapors below him, stretching out to the ocean, while the sky above was unclouded, the full moon illuminating the "ethereal vault." All was silent, save for a breach in the mist, a blue chasm not far off, a "breathing-place" whence came a "roar of waters, torrents, streams / Innumerable, roaring with one voice!" heard over the whole earth and sea and seemingly felt by the starry heavens.

When the scene dissolved and the poet thought about what he had seen, it seemed to him to be an image,

> the type
> Of a majestic intellect, its acts
> And its possessions, what it has and craves,
> What in itself it is, and would become.

The moon hanging over the mists, the light above, the sound below, the dark abyss and the silent sky—"There I beheld," the poet writes, "the emblem of a mind." The poet spins out his image of the mind, the relations of its parts represented by the moon and the waters, its powers resembling those of human imagination, and exerting profound influence on its thoughts and creations. The mind's creations can possess such mastery, he asserts, that they can catch even the creators by surprise,

> Like angels stopped upon the wing by sound
> Of harmony from Heaven's remotest spheres.

THE JOURNEY of the poet, the laborious climb through darkness and silence, should be familiar to anyone who has attempted to understand what seems first clouded in mist: the discomfort of the still air, the awkward pace on narrow paths, the isolated broodings of the climb. This book promises no less, but it hopes to provide something more, some hint of brightening by journey's end, some vision of the expanse and vistas that have opened to those who have made such journeys their lifework, some inkling of the powers and forms that compose these emblems of mind.

There are two paths to be negotiated here—each with its own twists and treacherous turns, each with its separate maps and resting places. The paths are those of music and mathematics, and the claim that they are similar, or at the very least related, has become a commonplace—as has the claim for the vast illumination they offer to those who pledge themselves to the climb. But it is a commonplace shrouded in mystery. Connections between the two have had almost no importance for the development either of math or of music; mostly any relationships have been irrelevant to their practitioners and creators, and mystically vague to everyone else. They are simply accepted without explanation or discussion, without even realizing what an unlikely pairing these two arts are. Why should there be any links at all? What do these activities really share? Do they share meanings or techniques or ideas? Why should we even suspect that they lead to a similar destination, let alone reveal similar visions? And if they are similar, is it simply because they are alike in the way all human creations are?

Music, after all, is amorphous: it shifts its texture and character from place to place and time to time. It can be crystalline or cloudy, sentimental or bombastic. It is transitory: when played it dissolves into memory. Mathematics, by contrast, is straightforward: it never alters its character, and it seems to soar above both place and time. Music is in the fray of things, played on grass reeds or gut strings, through brass tubes or hollowed bamboo, using all kinds of materials, natural and otherwise. Mathematics is, above all, spun from abstraction, not even requiring pencil and paper, which only record thought the way a tape recorder records music.

Unlike mathematics, music seems useless. A world without music would still provide food and clothing and shelter and uncounted luxuries; aside from the absence of such awkward and mechanically tortured contraptions as the piano and the clarinet, that world would be physically identical to our own. Music's main function seems to be as accompaniment for shamans and magicians and the sales pitches of Muzak contractors. But mathematics has left no part of our world untouched. It is used in drawing property lines, building submarines, predicting the curvature of space, solving algebra problems ("If two men can paint a room in three hours, how long will it take for three men to paint a room?"), and routing city traffic. A world without mathematics would be utterly different from our own.

Music seems steeped in affect; we commonly talk about music as sad or happy or angry or gentle. Music is spiritual, aesthetic, religious. Mathematicians couldn't care less about the emotions suggested by a theorem's proof. Has anybody ever encountered a "sad" theorem, or presented an "angry" proof, or inspired a courtship through abstract musings about topological spaces?

Mathematicians insist they are concerned only with the true, which under their glinting eyes increases in quantity with each generation. A problem unsolved in one generation is penetrated in the next; issues which gave Descartes sleepless nights are now tackled by high school students for homework. Music is another story. In what way is Stockhausen a step forward compared with Bach? When we learn a piece of music, what more do we learn other than the music itself? Musical knowledge, if it exists at all, is peculiar—incomprehensible from culture to culture, barely recognizable from time to time. For example, the sound of Greek music is lost to us completely; Gregorian chant seems unrelated to the Classical sonata. Music seems closer to language in its multiplicity and mystery than to mathematics; it is rooted in individual cultures. Play a South Sea fishing chant in Carnegie Hall, or a Beethoven piano sonata in the Australian bush, and you mix media and meanings; something is altered.

By contrast, mathematical truth stands as a rebuff to music and any of the arts: it seems as untied to space as to time. The Babylonians may have had a counting system based on the number 60 (whence comes our eccentric time measurements for minutes and seconds), but they reached the same results we do today whether we use a system based on the number 10 or, in our computers, a system based on the number 2. The prodigious mathematician Srinivasa Ramanujan could, in an isolated hut in India, teach himself mathematics from two mediocre textbooks, then send his brilliant results off to Cambridge, because what is true in one country's mathematical work is true in another. Mathematics seems free from cultural influence or constriction.

So can there be two more alien subjects with which to cultivate poetic metaphors? Even painting and music might be more likely mates. For the distinctions between math and music seem fundamental: they are between truth and beauty, time-lessness and change, science and art.

But these distinctions are less fundamental than they seem at first. We need to step back and examine the notion of comparison itself, the ways in which we examine like and unlike things. Things can be different in so many ways; how do we know which ways are significant? A person is like a cabbage because both need air, and unlike a mackerel in choice of habitat; is a person then more like a cabbage than a mackerel? Whenever we compare two seemingly different things, we are caught in a web of questions about our meanings and our intentions, about what we consider important and why.

So before setting out, we must understand something about the ways in which we understand similarities. The most obvious way we show things are connected is to give them the same word: when children learn language, they learn concepts, ab-stractions that are embodied in a name. A nose is not a particular nose but a general one; it can even be recognized as a "nose" when almost all distinguishing characteristics, like shape and nostrils, are removed aside from its place on something recognized as a "face." A toy car is called "car" just as a real one is, even though they may share neither size nor shape, design or color. These notions are learned and applied in an utterly unselfconscious way. When a connection is made, it immediately specifies what is important, and what incidental. It is unimportant, for example, that a toy car have doors or small seats or even a windshield to be recognized as a car by a child; what is important is that it have four wheels and a bulge on top, and that it move. Making connections requires some sense of the essential; it requires abstraction.

When identifications and links are newly made, when they no longer depend on the given abstractions of language, then the project becomes still more difficult. We can no longer rely on words which represent connections previously confirmed. For instance, when we want to discuss how a car is like a doughnut, we need to find aspects of a car and a doughnut that are shared and give those characteristics some other name. When the notions are still more general, the challenge is daunting. In what way, to take a cue from Lewis Carroll, is a raven like a writing desk? Creating an abstract link between abstract concepts means understanding each thoroughly in itself, then understanding their relationship to each other. Thus, in comparing things so obviously different as mathematics and music, we must consider the essential aspects of each. We must abstract from mathematics and music their practices and concerns, discern their inner lives, then tease out any signs of shared patterns. We need to define what mathematicians call a mapping between one world and the other.

This is something we do any time we compare, even if the objects or concepts under consideration seem at first to have nothing in common. For example, we know that a raven has two feet and a writing desk none, that a raven is alive and a writing desk is not, that a raven flies and a desk sits. But we can also look at the raven and the desk in other ways. We can see that a raven's feet steady it on the ground and its wings steady it in the air, and that that natural stability is imitated by the manmade desk, which is meant to neither tip nor topple under pressure. We might even argue that a writing desk has a certain function decreed by its maker, and take a religious perspective for arguing the same about the raven. Or we may take the raven's symbolic and associative significance of darkness and foreboding, and connect that with the associations called to mind by any writer facing his desk preparing to write a first draft ("Quoth the raven . . ."). The point is that a raven may indeed be like a writing desk, if our interests lie in certain directions. It depends on the context in which we look at them, on the domain in which we wish to compare them, and on what we consider important.

So too with mathematics and music. The links between them may lead us into profound regions we would never have stumbled on if our path were guided solely by one or the other; and our understanding of mathematics and music is bound to change based upon those connections. We may even come to see that the process by which we reason about them bears an uncanny resemblance to the processes at work within both—but that is anticipating what may lie at the end of a still tangled path. The beginnings are not in the gross categories of art and science and beauty and truth but in detail, in attitudes and approaches, in the human activities undertaken.

WHAT EXACTLY is "doing mathematics" or "making music"? Mathematics does not explore the structure of the physical universe as does physics or biology; it is something else entirely. But what? That simple question has puzzled philosophers of mathematics for several millennia. The working mathematician tends, generally, to avoid such questions about mathematics, just as most of us do about daily life. The mathematician works on problems, teaches students, and proceeds much as any other professional might. The musician does the same, caught up in the learning of repertoire or the composing of new works and the teaching of students but rarely approaching questions about what music itself might be or what its role is in culture. Music for the musician is like math for the mathematician: it simply presents itself as a mystery to be worked with and around. This is also why the philosophy of music has remained fairly primitive; what cannot be explained is passed over in silence.

Consider the most basic experience of "making" music. When I set out to learn to play Bach's D♯ Minor Fugue (from Book 1 of *The Well-Tempered Clavier*) on the piano, the music seemed frustratingly intricate. There is a theme that is simple enough: it begins with a leap upward, but it is felt less as a leap than an unfolding. It should be heard as if the second note grows out of the first, opposing it but also connected to it. The theme then turns with a plaintive caress and, as if taking a breath, gently echoes its own beginning before sadly returning, step by step, to its origins. The gesture's two parts have almost different characters—an excursion and a return—but the commanding spirit is melancholic, unsettled. When the theme reenters in another fugal voice, the restatement rises out of its own lingering sigh.

The problem in learning to play the fugue was not just training the fingers to create this voice so that it seemed to grow out of itself; it was to have each voice create a seamless line while two other such lines were proceeding at the same time. With practice, that could be done, the fingers passing notes among themselves, sliding one to the other, all the while keeping each voice intact. The problem was to hear these voices at the same time, to sense, when one voice entered with all its notes quickened, and another entered with all its notes slowed, and a third playfully combined the two timings; to play these properly required being able to focus attention above them all, so that at any instant any of these voices could be followed, while the integrity of the others was unaffected. One trick to ensure that no voice was slighted was to sing one voice while playing the others. This was not just a matter of learning to hear; it was a matter of comprehending, anticipating, so that the different layers of contrasting sound became strands in an integrated texture. Just as the theme needed to be felt as a single gesture rather than as a combination of notes and intervals, its combinations and varied statements needed to be bound together into a dramatic meditation, until the entire fugue could be seemingly exhaled in a single breath, its machinations becoming manifestations of natural forces, the melancholic theme building gradually to grandeur.

The learning of a composition combines physical, aural, and intellectual work, each feeding the others. Mathematics may leave out the physical and aural, but the intellectual work is daunting. At about the same time these labors with the Bach fugue were in progress, I was engaged in a more large-scale exercise. It involved not a single theme with a determined character but many disparate voices, each confusing in its own way. These voices had names. They were disciplines, regions of study: they were called analysis, algebra, set theory, and topology. Each one seemed to set up its own laws and then follow these rules wherever they led, spinning out theories of curves or numbers or spaces or objects which had no obvious meaning at all. My labor of understanding was divided here, attempting, for example, to comprehend how a theorem about measuring area under a curve was related to a theory of probability, then trying to understand why the notion that we can always choose individual objects from collections of objects is a mathematical notion—indeed, that it is one of the basic facts in an entire area of mathematical research.

I didn't even know until later that there was anything uniting these ideas. I was working through problems, trying to comprehend details; it was like training my fingers to play the right notes, learning the ropes—nothing more. But then, as when a pianist is gradually getting patterns into fingers and mind, here and there, intermittently, within each of these areas, there would come a time when seemingly different ideas would become linked, patterns would appear. Suddenly there were connections where none had seemed to exist before, abstract systems seeming to mirror each other, or at least pass through each other before going their separate ways. The mind, like the ear, was set free.

These moments of illumination—finding relationships and hearing links where none existed before—are known to all who have been students of mathematics and music, but it would be hard to say just what is learned or how. That illumination also seems to have little relationship to life experience. In both disciplines the extent of one's understanding seems strongly determined by "gifts"—gifts of temperament and insight, ability to coordinate concepts (and, in music, to coordinate fingers and breath with concepts). The "prodigy" is a figure almost native to music, mathematics, and other activities like chess—dependent less upon experience with the world than with insight into a seemingly closed, abstract universe. For example, Felix Mendelssohn had, by the age of seventeen, composed a dozen symphonies and the famous *Midsummer Night's Dream* overture. Mozart's abilities at an even earlier age caused him to be paraded around Europe by his ambitious father. Despite early abilities, of course, musicians may require experience to mature into great artists; but mathematicians do not require even that. The mathematician Carl Friedrich Gauss corrected his father's calculations before he was three years old and jested that he knew arithmetic before he could speak. Mathematics is a "young man's game" as the mathematician G. H. Hardy put it (and few enough women have gone into it). Hardy pointed out that Galois died at twenty-one, Abel at twenty-seven, Ramanujan at thirty-three, Riemann at forty—all having achieved near immortal stature in the history of mathematics.

This is not to slight the genius of lifetime practitioners of music and mathematics: they have the positions in their worlds of true visionaries. In music this is how we view Beethoven in his late years, like a Newton, voyaging in strange seas of thought, alone. Einstein's halo of hair proclaims his otherworldly authority—as mathematician and seer—with as much emphasis as do the theories that bear his name. Palestrina, Bach, Wagner—the names strike the same awe into musicians that mathematicians find in the names of Gauss, Cantor, von Neumann. Tales are passed around about all manner of accomplishments: pianists who can read an orchestral score at sight at the keyboard, conductors who can hear a wrong note in the midst of scored cacophony, composers who hear without the need of ears, mathematicians who blithely toss off answers to problems that had kept colleagues baffled for months or who see that the entirely wrong questions have been asked. These are more than magicians. In mathematics and music the possessors of genius become members of pantheons, their glory standing apart from the achievements of other great artists, whose labors are tied to ordinary life. Those who are elect are often both blessed and cursed: unsuited for ordinary existence, they are often socially brutish—like Beethoven—while being apparently touched by the gods.

Math and music are both so abstract they can seem otherworldly, but both also have extraordinary this-worldly power—music in its effects on the listener, mathematics in its applications in the world. Both activities are fully understandable only to the elect, those who have been granted the necessary vision and passion and who have expended the necessary labors, but in their application and performance they are accessible to all. It is not inappropriate to speak of the "calling" of mathematicians and musicians. This image is not casually religious; mathematics and music have been intimately connected with ritual and revelation; they have long been associated with mystery and risk as well.

Because of the mixture of the esoteric and the practical, the otherworldly and the temporal, the universal and the particular, music and mathematics have held almost mystical power since ancient times. Histories of the early years are sketchy, but the power of those who held the secrets is clear: it extended over the earth and the heavens. Herodotus wrote that geometry began in Egypt out of the need for surveying after the annual flooding of the Nile; Aristotle argued instead that Egyptian mathematics grew out of the speculations of the priestly leisure class. Aristotle was probably projecting the Greek experience onto an earlier culture, for it was in Greece that the theory of mathematics came to its first maturity, taking primacy over utility. There was, however, no rigid separation between the intellectual play with numbers and their uses in the observation of the heavens, the surveying of land, the construction of buildings, or, as the Bible tells us, of temples.

Music was, by its nature, most probably the possession of all people and existed even at the most primitive level, but its highest systematic and rational cultivation must also have been reserved for the priestly classes, among whom it became part of ritual—and so possessed a special place in early conceptions of the universe. In fact, the need for specialized knowledge in the practice of mathematics and the practice of music may have conferred on both a prestigious position. There must have been then, as there are now, castes of knowledge, culminating in priestly understanding: this was, of course, the case with the music of the Catholic Church in the Middle Ages. The Pythagorean order is the most famous of those based on esoteric knowledge. It was communal, secret, subscribing to the doctrine of the transmigration of souls; it is said that Pythagoras himself invented the Greek words for philosophy and mathematics.

The magical and ritualistic aspects of mathematics and music are evident even today when we pick up any score or open any math text. We are faced, in either case, with a script different from the scripts and ideograms which are symbols of spoken language. Mathematics contains words from ordinary language mixed with signs that have no simple translation—symbols like $\int$ or π or $\emptyset$—along with combinations more exotic:

$$|f(0)| \prod_{n=1}^{n(r)} \frac{r}{|\alpha_n|} = \exp\left\{\frac{1}{2\pi} \int_{-\pi}^{\pi} \log|f(re^{i\theta})| \, d\theta\right\}$$

The result is a language that can be neither read nor understood without initiation. For example, what are we to make of this sequence of sentences from a math text:

> Let Γ^* be a nonnegative cochain complex of fine presheaves of modules on a paracompact Hausdorff space X. Assume that for some integers $0 \leq m < n$, $H^q(\Gamma^*)$ is locally zero for $q < m$ and $m < q < n$. Then there are functorial isomorphisms:
>
> $$\check{H}^{q-m}(X; H^m(\Gamma^*)) \approx H^q(\hat{\Gamma}^*(X)) \qquad q < n$$
>
> and a functorial monomorphism:
>
> $$\check{H}^{n-m}(X; H^m(\Gamma^*)) \to H^n(\hat{\Gamma}^*(X)).$$

We may recognize many of these words, see a few references to numbers, and know that other words must possess special definitions within mathematics, but the symbolic expressions are condensations of reasoning that have taken pages of earlier text. Reading mathematics is not at all a linear experience; the eyes of even the most accomplished mathematician reading advanced material rarely move from left to right across the page. Understanding the text, even for a virtuoso, requires cross references, pauses, scanning, and revisiting. The meaning is rarely completely transparent, because every symbol or word already represents an extraordinary condensation of concept and reference. Of course, reading even ordinary language is complicated, but reading mathematics presents a different order of complexity; it involves a return to the thinking that went into the writing.

When we look at a musical score, we are still further from ordinary language. Readers of Italian and German will find some familiar words scattered here and there (allegro, klagend), but, unlike in ordinary writing, in music there is a deliberate sense of space being used and defined: vertical space punctuated by rows of five-line staves, themselves arrayed with black strokes, and horizontal space measured by vertical lines across the staves.

Here, too, reading is governed by something other than the mere concatenation of symbols; it is measured by the tempo indicated at the start by words (allegro con brio) and, often, metronome markings ($\flat = 108$). We "read" according to the passage of time, imagining sound. It also is not purely linear: it involves the vertical dimension along with the horizontal, the first presenting a form of musical space, the second the progression of musical time.

Neither mathematical nor musical script is easily translated. And while the ordinary written word can be an end in its own right, neither of these scripts is. They are records of thinking and hearing. Mathematical script is so compressed that to be understood it must be unwound, translated into the reasoning that took place at its birth. Musical script is only an approximation, signs meant to point directly to sound itself. Neither the mathematical script nor the musical script is the main point. Both point to universes quite different from the one in which ordinary language functions so well.

Opening of Beethoven's Fifth Symphony

But in each, too, there is a genius in the very notation that has developed for giving representation to ideas that seem to lie beyond ordinary language. Until the development of the zero as a place holder, aspects of arithmetic remained obscure; with the zero, the manipulation of symbols can itself provide revelations. We know something profound about the sum of 100 and 10 when we write it as 110 which we could not know if we were to write it in Roman numerals as CX. In advanced mathematics the first step to understanding is developing a notation to give concepts a structure and sense. There are times, in fact, in puzzling through a problem in mathematics, when similarities in notation are the first clues to deeper relationships.

Similarly, the musical stave not only created a structure within which Western music could develop but also shows something other than just the sounds being made. It indicates how the various elements stand in relation to one another, how sound creates a space. It gives a pictorial representation of pitch, mapping out the high and low; it shows how different musical voices move against and through one another; it links the dimensions of pitch and time, and gives image to rhythm. The notation allowed harmony and counterpoint to develop as they did; it influenced the course of the art it proposed mutely to serve.

The power of the scripts is partly the power of the arts; we manipulate one to discover the other. The scripts have an uncanny relationship to their respective arts. A resemblance in script can reveal a resemblance in substance. And the act of creating scripts is almost inseparable from the act of advancing the art. The scripts take on lives of their own. They are so laden with meanings and implications that they become iconic. A floating eighth note or a scrawled integral sign represents the art itself.

The two scripts also reveal something about the character of the two arts, for representation and resemblance are at the heart of music and of mathematics. This is one reason for their ritualistic character. Magic and ritual often require actions on objects that affect other, distant objects. Generally the effect is created through similarity of some kind. The similarity might be in the character of an object—a candle might represent more general powers of heat or illumination. The resemblance might also arise through a physical connection involving an object that "belongs to" or is connected to another—the way a crown can become endowed with the powers of royalty. Magic and ritual require metaphor and metonymy; they require a poetic imagination.

Some medieval Jewish mystics, for example, believed that we could create a metaphorical diagram of the godhead that resembled a man's body; to each part of this body was assigned an attribute of the divine. This metaphor also meant that there was a link between the parts of the body of a man and the parts of the "body" of God; actions of one could have an effect on the other. In this way a metaphor became literal. "The arm" of God could be magically affected by using the arm of man. "Metaphors for us, but not for Him" is the way one mystic put it. Prayer—consisting of formulas of words and letters—worked its effect through similarity as well. There was a system, gematria, which suggested that the letters of the Hebrew alphabet were connected to essential properties of the universe. Each letter was assigned a numerical value, which could then become a magical link between seemingly unrelated words: if their letters added up to the same number, it revealed a deep connection between the words and between the objects they referred to.

This is related to our earlier thinking about finding the similarity between seemingly unrelated objects. In a religious context, similarity is the foundation of prayer and ritual, bringing together the earthly and the eternal. Because music and mathematics seem to participate in both realms, they have been used, like the gematria of the Jewish mystics, to control and connect them—realms ranging from the internal world of the music listener to the stars in the heavens. The genius is both seer and priest because he can make visible the hidden connections within music and mathematics that reveal, as in the mystics' diagram, hidden connections among outside phenomena.

The Greeks were quite explicit about this. History may merge with legend concerning the Pythagoreans, but their doctrines had extraordinary influence. Mathematics was accompanied by number mysticism, with each integer having a metaphysical significance: the number 7 was linked to the goddess Athene, 5 suggested marriage because of its combination of the first even with the first odd number greater than one. Aristotle noted that the Pythagoreans treated the principles of mathematics as the principles governing all things. This went along with a mysticism about music, which was itself linked to number and to the character of emotional and intellectual life. These deep relationships were sensed by other Greek thinkers too. Plato rejected musical styles that corresponded to undesirable states of the soul—the plaintive or effeminate, for example—and, in *The Republic*, praised only the Dorian and Phrygian modes, which are associated with courage and sobriety. Aristotle explicitly argued in his *Politics* that rhythms and melodies are sonic translations and representations of moral qualities. These musical connections with the state of the mind and soul were connected, as the Pythagoreans asserted, to the harmony of the universe as well.

Two thousand years later, Johannes Kepler, in publishing his laws about the movement of the planets, made the religious claim for mathematical exploration explicit: "The die is cast; I have written my book; it will be read either in the present age or by posterity, it matters not which; it may well await a reader, since God has waited six thousand years for an interpreter of his words." In the seventeenth century, Gottfried Leibniz, at the age of twenty, thought he could create "a general method in which all truths of reason would be reduced to a kind of calculation. At the same time this would be a sort of universal language or script, but infinitely different from all those projected hitherto; for the symbols and even the words in it would direct reason." Such a language would display the foundations of similarity; it would be a script that would contain revelations, all truths and no falsehoods. It would be the language of God's scripture.

The claims for music have been similar, its powers reaching within, to the soul, and without, to the heavens. John of Salisbury, bishop of Chartres from 1176 to 1178, had observed, "The soul is said to be composed of musical consonances" and through the laws of musical proportion "the heavenly spheres are harmonized and the cosmos governed, as well as man." Four hundred years later, Kepler claimed his laws revealed a music of the spheres—the movement of the planets creating tones as they passed through the heavens.

In testimony to these links, musicians and mathematicians often seem to be exchanging positions. Jean-Philippe Rameau began his classic treatise on harmony in 1722 by praising mathematics ("I must confess that only with the aid of mathematics did my ideas become clear and did light replace a certain obscurity"). Galileo in the seventeenth century speculated on "why some combinations of tones are more pleasing than others." The eighteenth-century mathematician Leonhard Euler wrote a discourse on consonance. The development of modern analysis—the area of mathematics that expands on the abstractions of calculus— was inspired in part by attempts to describe the motion of vibrating strings. Twentieth-century music has seen a radical mathematicization of musical language, with invocations of set theory, Markov chains, fractals, and assorted other seemingly irrelevant subsets of mathematical discourse. The divinity of these relations is often taken for granted. Claude Lévi-Strauss, whose structural analysis of myth uses the vocabulary and techniques of mathematics, wrote in *The Raw and the Cooked*, "The musical creator is a being comparable to the gods, and music itself the supreme mystery of the science of man, a mystery that all the various disciplines come up against."

The nineteenth-century mathematician James Joseph Sylvester gave mathematics the same exalted status Lévi-Strauss gave music. He was a lawyer, an actuary, an accomplished musical amateur who took singing lessons from Gounod and was said to have been immensely proud of his high C's. But his passion was reserved for mathematics, and his hardheaded scholarship was punctuated with breathless evocations. Mathematics, he wrote, "is limitless as that space which it finds too narrow for its aspirations; its possibilities are as infinite as the worlds which are forever crowding in and multiplying upon the astronomer's gaze; it is as incapable of being restricted within assigned boundaries or being reduced to definitions of permanent validity, as the consciousness, the life, which seems to slumber in each monad, in every atom of matter, in each leaf and bud and cell, and is forever ready to burst forth into new forms of vegetable and animal existence."

BUT TO UNDERSTAND the reasons for such effusive glosses on these arts, we must turn, briefly and tentatively, to their humble origins. For music, they can be found in the simple vibrations of simple strings. We can give such vibrations a less-than-mystical concreteness: gently extend a telephone wire and start to swing it right and left in a regular rhythm. Between the two fixed ends of the string—one at the wall, one held firmly in the hand—we can see a wave form swelling toward the center. If the cord is shorter, we have to swing it faster to create that single wave; if it is longer, we must move more slowly. In fact, double the length and we halve our movements; halve the length and we double our movements. Each length of cord seems to have a natural rhythm associated with it. This was Pythagoras's law.

That rhythm might also be described in terms of period or frequency (we swing it three times a second perhaps). But if the vibrations are fast enough, we begin to *hear* it. The vibrations create a note, a pitch. And just as each telephone cord has its own fundamental frequency, moving simply between the two

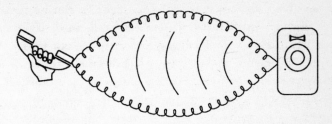

A swinging telephone wire in vibration. Each length of wire has a distinct rate of swing.

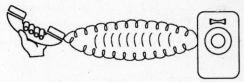

A telephone wire of half the length will swing twice as fast.

fixed ends, each string has its own fundamental pitch. With only a few qualifications we can start finding fundamental pitches in all kinds of objects: tree trunks swayed by the wind or columns of air in soda bottles; some have insisted, in fact, that similar fundamental pitches exist for the human soul or the revolution of the planets about the sun. In this way, sounds and numbers have become intimately related. And their connection is not arbitrary. It is not a metaphor: if we interpret the words properly, sound is simply heard number; number is latent sound.

Let us stay with that swinging cord, now moving between two fixed points. If we gradually increase the rate at which we let it sway, at first the wave breaks up. The string seems to jerk randomly; it becomes difficult even to hold. While the initial wave let our hand stay steady, suddenly the string pulls against us. There is a rebellion against our determination to fix the string's length and position. At a particular moment, though, a sort of vibrational harmony is established. The struggle ends.

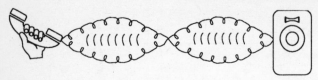

Double the rate of swing and the wire vibrates in two halves. If the wire sounded a pitch of middle C in the illustration on page 23, this one would sound a C an octave higher.

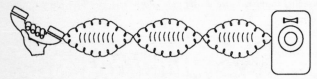

And this would sound an even higher G.

Our hand remains fixed. This happens when we are letting the cord swing at precisely twice its previous rate.

Twice the rate, according to Pythagoras's law, requires half the length, and, in fact, the string then begins to vibrate in separate halves with a fixed point in its center and a fixed point at either end. The same thing happens as we increase the rate again, until there are three segments, then four. The string must always divide into equal parts, with equal pitches associated with each part. If we could listen to our vibrating cord, we would start to hear a series of pitches, an arpeggiated chord. They would begin, perhaps, with C, which would be the string's fundamental note. The division of the string into two parts would create another C, just an octave higher. Then things would get more interesting. With three divisions we would hear a G, a note that musicians would say is a fifth higher than the last C; with four divisions we would get a still higher C; with five would come the note E, a third higher than C. The notes would continue to climb, always returning to C at ever higher levels, while including other notes and intervals, with ever smaller intervals between them.

We could depict this phenomenon like this:

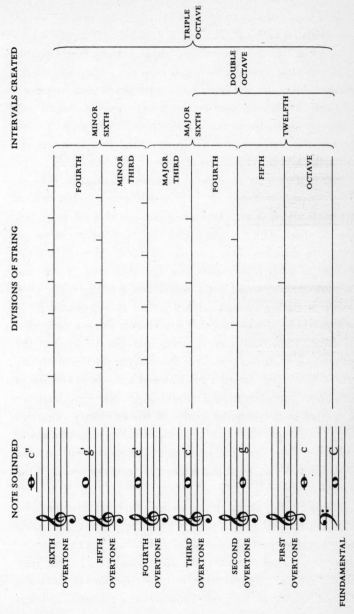

Overtones of a vibrating string. The prime signs (C'') mean the same note is heard in higher octaves.

If we were to listen to this ever climbing sequence, we would find something unusual. The initial gaps between notes would be greater (C to C as opposed to, perhaps, C to E), but the notes themselves would *sound* closer. It is even easy, when singing, to confuse octaves, to find one's vocal cords doubling a frequency or halving it without noticing. Even between the second and third overtones—the G and the C—there seems to be a close relation. The harmony we felt in our hand is a harmony we hear with the ear. This is the experience of consonance.

In fact, if we were to sound at least the first six such divisions of the string simultaneously, we would hear a sound that at least in the West is very familiar. It consists of a three-pitched chord, a triad, which is the foundation of Western harmony. And if we sounded these pitches simultaneously, so that the intensity of each higher note was less than that of the one preceding it, we would get a sound that we might not even identify as having distinct pitches within it. We would get a single rich string sound, one that would seem to have dimension and depth. Remarkably, every string that vibrates even at the fundamental pitch also vibrates at these higher pitches in varying degrees. If we only heard a fundamental tone, it would sound artificial, so "pure" as to be without body. Timbre and intricacy are created in instrumental sounds by the subsidiary vibrations that cascade along the vibrating string, dividing it first into two, then three, four, five, and so on until the movements become insignificant. These are the overtones, elements of sound in the physical world.

The overtones create this sequence of pitches, but they also create unexpected relationships within the pitches. If a string vibrates in three parts, for example, we hear not just the relationship of the second overtone with the fundamental but the relationship between pitches created by higher divisions of the string. We start hearing unexpected resonances and relations, all latent in a single tone. The scale, Arnold Schoenberg observed in 1911, in his famous text on harmony, is the "analysis" of a single tone, since the overtones of a single tone will eventually generate the most important distinct pitches. The chord, he said, is a "synthesis" of tone, the collapsing of its elements into a single sound, in which overtones combine to create harmony. Indeed, one fact of physics is that most sustained sounds can be decomposed into a series of overtones like the ones we have been describing. So scale and chord are, in a sense, compressed within individual sounds.

Every culture uses these physical facts as foundations for its musical systems; choices are made about where to put emphasis, how to define degrees of consonance or dissonance, the permissible and the forbidden. In the West, the sequence of overtones was interpreted in different ways; only toward the end of the fifteenth century was the third considered a legitimate consonance (there were claims made for the fourth because it is the inverse image of the fifth). But also in the West, partly because of the influence of Pythagorean mysticism, the relationship between the sounds and numbers in this overtone series became central. For about three or four hundred years—at least until the twentieth century—this relationship took on the quality of scripture. For the initial overtones creating the triad chord have another implication. Remember: the triad is composed of the C, E, and G of the first six overtones. The fundamental tone and the fifth—the C and G—have a very close consonant relation, but there is also a tension between them. For if we create a string vibrating with a fundamental of G, its *own* strongest overtones do not overlap closely with the overtones of C:

overtones of C: C′ G′ C″ E″ G″
overtones of G: G′ D″ G″ B″ D‴

These consonant tones then, have strongly *dissonant* overtones. Sound a G and a C together, and in the midst of the pleasing congruence of like minds, we might hear high D's and B's jarringly framing the original C.

In Western harmony, these two triads, on the fundamental and the fifth, are the poles of tonal musical drama. They are called the tonic and the dominant; the strongest overtones of the dominant rub against the tonic but also seek to meet it. And since each tonic is also the dominant of another tonic, we get a series of tensions and relations, in which chords begin to take on character in relation to others. We get an architecture of musical space. These chords are numbered, subscripted, super-scripted, and inverted. They become ingredients in a subtle calculus of relations. Under musical scores, analysts scrawl the numerical logic: I, IV, V_7, II♭, VI, i, I. The most obvious mathematical aspect of music is simply its systematic organization out of the physical facts of vibrating strings or air columns. Western tonality almost rivals the ordering of a mathematical system; twentieth-century substitutes have become even more complex.

Only recently has there come to seem something arbitrary about this systematic ordering of sound; every culture previously had something resembling natural law to govern the ritualistic combination of pitches and rhythms. This relativity is, no doubt, a source of much opportunity. But it is also distracting; it clouds the origins of the arts. To see more deeply, we must begin to put aside such notions, at least for the beginning of this journey.

WE MUST take number and sound as seriously as they once were taken—as, say, by the Greeks: when they discovered that numbers exist which are neither integers nor ratios of integers—numbers which confounded all their notions of harmony and rationality—they were so horrified that the discovery was kept secret. *Alogon*—the unutterable—these numbers were called.

Proclus told that those who brought these numbers out of hiding perished in a shipwreck: "For the unutterable and the formless must needs be concealed. And those who uncovered and touched this image of life were instantly destroyed and shall remain forever exposed to the play of the eternal waves."

The irrational was forbidden because there was thought to be a direct link between number and the order of the soul and of the universe; a disruption in one was a disruption in the other. This notion of number was literally a notion of ratio— ratio understood as not just the quotient between two numbers but the expression of rational thought and understanding. The "unutterable" numbers—such as the square root of 2, the cube root of 20, the circumference of a circle with a diameter of 1— are still called irrational. No matter how much time we spend calculating them, no matter how far we proceed, we can never express them exactly. We shall return to this idea later; for now, we must just accept that these numbers can never be made "rational."

There are, of course, musical counterparts. Centuries after the Greeks, the Western church treated one interval in just this way: the tritone (the leap from C to F♯, for example) was found to be "unutterable." It was forbidden and called the *diabolus in musica*, the devil in music. It was harsh, unresolvable, dissonant; it split the octave down the middle. It stood outside the theoretical musical systems that had been constructed; it could not be accounted for in the combinatorial musical systems of the Renaissance. Its use guaranteed the equivalent of spiritual shipwreck; during the nineteenth century, whenever Faustian bargains were invoked in music, the demonic tritone appeared.

But this metaphorical similarity between the irrational and the strongly dissonant is more profound and more mysterious than we have any right to expect. A musical interval and a mathematical number have been judged similarly: they have been considered beyond the limit of what is permissible. They have been sensed to be not just unpleasant but dangerous to an established order, not just threatening but demonic. This must be better understood. But for now, perhaps, it may be enough to see that in both mathematics and music there have been notions of natural and unnatural. In both mathematics and music there have been notions of connection, linking the soul and the universe. In demonism and divinity, mathematics and music exhibit extraordinary affinities.

But what lies at the heart of this intimacy? What could the links between mathematics and music possibly be? Why does the imagery of the eternal and the divine, the natural and the unnatural, seem to hover so persistently around mathematics and music? If one claim is that there are similarities between them on the one hand and the structure of the soul and the structure of the universe on the other, in what could such similarities lie? How also are we to understand the "art" of mathematics and the "science" of music?

We can already see that the origins of harmony and ratio in the vibrating string have a numerical character, that a language of chords creates a system of order and law that can seem mathematical, but these are the most trivial relations, the simplest, most obvious, and perhaps ultimately least important. If we are to believe the assertions of similarity and connection, there is something that mathematics and music share with our notions of the universe and our notions of the mind and soul— and our notions of beauty as well.

Both music and mathematics create order, worlds in which processes occur, relationships are established, and elements are regulated. These worlds possess structures that might be mapped into our own; they might be similar in the strongest sense the mystics allow. True, mathematics and music go about constructing their worlds in different ways, and exploring of their similarities will require extraordinary caution. But through such an exploration we can begin to glimpse the sorts of order and truth they both create.

This will not be an easy task, for it will involve a certain amount of immersion in both music and mathematics (particularly in the next two chapters). For a time, we might even have to focus all our attention on detail, in the expectation that when we lift our eyes a grander shape will emerge. We must begin to find out how mathematics works, how it starts from the world and proceeds through abstraction. We must search for its musical spirit, the variations in mathematical style, the play with similarity and form, the connections and inventions created in abstract thought. We will follow a mathematical theme, trace its evolution, hint at some of its implications. Only then can we return to music and see similarities in how its deepest structures work, how its abstract arguments proceed. We must seek music's truly mathematical spirit, its rigorous establishment of similarity and transformation, its uncompromising engagement with our intellect. Only then can the questions we began with be returned to if not fully answered: how the image of the divine arose, how beauty emerged in science and art, and how, out of beauty, a power grew that is both mystical and concrete.

Sylvester wrote in a paper on Newton: "May not Music be described as the Mathematic of sense, Mathematic as Music of the reason? Thus the musician feels Mathematic, the mathematician thinks Music—Music the dream, Mathematic the working life— each to receive its consummation from the other when the human intelligence, elevated to its perfect type, shall shine forth glorified in some future Mozart-Dirichlet or Beethoven-Gauss." Sylvester's area of research in mathematics was the invariant— mathematical objects that remain unchanged when all is being transformed around them, objects which form a sort of kernel (a word that has a technical mathematical meaning matching its metaphorical one). It is not surprising, then, that he was sensitive to the ways in which things are compared, transformed into each other, how mathematics and music might interact. We must take his sort of comparison seriously. The metaphorical climb we are beginning must combine thinking and feeling, so that by the end we will stand, like the poet on Mt. Snowdon, gazing with wonder and respect at these emblems of our minds.

II

PARTITA:
THE INNER LIFE OF MATHEMATICS

Mathematics, as much as music or any other art, is one of the means by which we rise to a complete self-consciousness.

J.W.N. SULLIVAN

T HE START OF A CLIMB IS PROBABLY THE MOST DISCOURAGING. It requires from both guide and traveler confidence that the path is negotiable, that scattered debris will not cause undue hardship, that even detours may be of some importance in making the journey easier. And because the goal is not in sight, it is easy enough to feel lost, confused about whether progress is being made at all.

This is particularly true as we begin, for we are turning first not to music, which in its many incarnations has immediate and compelling power for all listeners, but to mathematics, which for most of us remains a nearly impenetrable mystery. Nearly all of us respond to music in some way, but we cannot all "listen" to mathematics. Some of our fear and confusion about this commonly applied but uncommonly understood art stems from the sad fact that the teaching and learning of mathematics are scandalously poor. The mathematics we all know—taught, at least at one time in America with some rigor—is a matter of solving problems, applying rules, learning the properties of geometric figures. It is, to a great extent, "applied" mathematics. We learn to construct a pentagon with a compass and straight-edge or to solve an equation for an unknown x. If we know the basic assumptions, follow the laws of logic, and are a bit clever, we can solve problems.

When we move beyond such questions, though, mathematics seems increasingly bewildering. It often seems to an outsider to be a collection of various esoteric techniques. Once trained in this specialty we are able to leap about with some agility, like a pianist perhaps, with virtuoso abilities. It is possible to manipulate symbols with ease, prove the Pythagorean theorem, multiply enormous numbers using logarithms, or find the focus of a parabola. But to what end are these various trills and roulades put? They make it seem as if mathematics is little more than a sophisticated tool, useful for specialists and hobbyists. If we are to proceed further, we must first take a step back from the mathematics we might be familiar with, away from the elaborate techniques and varied applications. We must turn to the ways in which mathematical knowledge is organized and the ways in which mathematical thinking takes place.

One place to begin is with a set of procedures which also might be familiar from high school geometry: the axiomatic method. It is based on the approach used in Euclid's *Elements*—which, aside from the Bible, has been cited as the most repro-duced and influential book in the history of the Western world. The axiomatic method gives mathematics the quality of a crystal-line mechanism: using it, we always know exactly where we are, where we are going, and how we are moving from place to place. The proof in classical geometry begins from premises and proceeds step by step, each justified according to statements previously accepted or proven.

The axiomatic method can also lead us closer to the inner life of mathematics, if we attend carefully to its nuances. In an axiomatic system knowledge begins with undefined terms and accepted postulates or axioms that describe the basic principles of the universe being explored. In Euclidean geometry, words such as *points* and *lines* are undefined terms, gaining meaning by appealing to our intuition. Axioms, also meant to be taken as indubitable, include such seemingly obvious "facts" as "Every line is determined by two distinct points" and "A circle may be drawn with any point as center and any given radius." We can then begin to find out what sort of universe is determined by these axioms, define different kinds of geometric figures, and explore their properties, all the while never stepping out of the bounds created by the axioms. Each proven theorem builds on another, each accumulating power through the seemingly immutable laws of logic. With enough time and energy, presumably every true statement would be discovered and proven, and no contradictions could ever be reached.

A formal axiomatic mathematics bears a vague resemblance to the music that we might imagine from studying texts in music theory. The overtone series leads to the laws governing the sequence of chords in the tonal musical language and to restrictions placed on simultaneous soundings of musical voices. Out of these axioms grows a musical grammar by which musical statements can be produced.

The only problem is that we know that neither axiomatic music nor axiomatic mathematics is a complete or satisfactory description of these arts. Music and mathematics are not "produced" or "generated" but come into being through much less rigorous and more fundamentally mysterious processes. There is something inadequate about these axiomatic visions. In music, we learn again and again how supposedly immutable laws and principles of form and organization can be violated, how new musical systems can be developed out of seemingly identical sonic facts, how widely varied cultures can create widely varied musics. In mathematics, the assumptions that axiomatics can determine the complete truth and that axioms have a necessary connection with physical reality have also been steadily undermined during the course of the nineteenth and twentieth centuries (about the same time that the tonal foundation of Western musical language was being questioned).

In Euclid's system, for example, there is a troublesome axiom—the famed parallel postulate. In one formulation it reads: "Through any point outside a line, one and only one line can be drawn parallel to the given line." This axiom always seemed less obvious, more ornate than the others, as if it needed to be proven rather than simply accepted. But all attempts to derive it from the others failed—and for good reasons. This axiom turns out to be independent of the others. We can accept all Euclidean axioms while contradicting that postulate, saying, for example, that through a point outside a line an *infinite* number of lines might be drawn parallel to it. Properly reinterpreted, this new axiomatic system actually creates a different geometric universe, one that is "non-Euclidean." We can begin discovering new laws governing that alternate world, such as the fact that the sum of the angles of a triangle is more than 180 degrees.

It took a long time for mathematicians to realize that such seemingly absurd notions might describe a viable space. This realization required treating the undefined terms in the Euclidean axiomatic system not as absolutes but as variables—abstractions that may not necessarily mean what we always have assumed them to mean. We can give them varied interpretations. On the surface of a sphere, for example, the word *line* can be interpreted to mean "great circle," a line around the sphere that divides it into two equal parts; in such a space the parallel axiom really does not hold—no two "lines" are parallel. Other interpretations create different spaces; if we interpret the Euclidean "plane" to be the enclosed space of a circle and "lines" to be chords in that circle, then an infinite number of "lines" can be drawn "parallel" to a given line through an outside point: within the "plane" they never intersect.

In the nineteenth century, the discovery that mathematics could be detached from the presumed requirements of the physical universe strengthened its abstract tendencies. An axiomatic system that had been immutable for millennia was suddenly reduced to an alternative, one system among many. The mathematical universe was no longer a mirror of the physical universe. Alternative mathematical truths, it was seen, could describe aspects of the real world, or they might describe nothing real at all. Mathematics became autonomous. Its meanings depended greatly on where it was being led and what interpretations were being given to its terms.

Mathematics, in this view, resembles a sophisticated machine, turning out strings of formulas without relying on confirmation from the world itself. A philosophy of mathematics known as formalism put forth this view with great rigor. It holds that all statements in mathematics can be "formalized," expressed as a string of abstract symbols which are then manipulated according to certain unchangeable rules. Mathematics, correctly monitored, would churn out "truth" after "truth." It would be like a player piano, perhaps, with the paper rolls prepunched, gradually revealing itself with each turn of the motor.

This view restores the absolute confidence once provided by the axiomatic method and actually raises the stakes. For formalism takes the idea that there are undefined terms on which a mathematical system is based and says that *all* terms in mathematics can remain undefined, that mathematics does not need to bear *any* relationship to the world, that it is only when its activity is concluded that we begin to put an interpretation to the strings of symbols and give them any meaning.

This mechanistic vision, of course, is entirely illusory: the dream of manufacturing all true statements through a kind of industrial application of rules and forms has proven to be just that—a dream. That was the import of Gödel's incompleteness theorem, which has become a cultural landmark as much as a mathematical one. The formalist image of mathematics we are naively heir to, as a kind of deterministic exercise accomplished according to strict laws of logic, was shattered. Formalism may have much to do with the way mathematics is taught; it may even have much to do with the way mathematics is popularly imagined; but it is fundamentally flawed. It simply cannot work.

It also has little to do with the way mathematics is actually done. The "doing" of mathematics—attempting to understand a universe that seems to be both invented and discovered by the mathematician—is no more dominated by compulsion or mechanism than musical composition is by the "need" to follow one type of chord with another. Actually the sensation of doing mathematics is, at times, an almost giddy one, of extravagant freedom, and frightening possibility. I was always amazed that the steady accumulation of theorems and proofs in the books I studied bore little relation to the way I thought about mathematics when attempting to solve a problem or prove a theorem on my own. This seemed to me a purely personal eccentricity (like my impatience with the descriptive power of standard musical analysis) until I realized that a three-line proof of a subtle theorem is the distillation of years of activity. It is not a picture of thinking; it is its final "formalized" draft. So understanding mathematics's inner life means understanding more than just the rules governing its formal systems; it means beginning to see why certain ideas are developed and others ignored, what is being searched for and how it is explored. We can no more come to understand mathematics through examining its final product than we can understand the experience of music through simply looking at a score or an analysis of one; there is an experience that lies underneath and behind the systematic organization of the material. This is by no means easy to explore.

. . .

As WE BEGIN AGAIN to look at mathematics to get some idea of that experience, consider a classic proof, one attributed to Euclid himself: that there are an infinite number of prime numbers. A prime number is a number not divisible by any other number (other than 1 and itself)—numbers such as 2, 3, 5, 7, 11, 13, 17, 19, and so on. Since a prime cannot be broken up into divisors, it is a sort of numerical atom. In fact, every number can be decomposed into a unique product of primes revealing its numerological and skeletal structure (e.g., $6 = 2 \times 3$, $360 = 2^3 \times 3^2 \times 5$). Showing that there are an infinite number of such primal primes, then, is crucial to our entire understanding of numbers. But how can something like this be shown? How can one demonstrate that there is no end to something?

The first temptation is to keep "making" primes, to see that there are always primes larger than the one just found. But there is no formula for primes, no way to predict where and when we have found one (computers have been used to test whether large numbers are indeed prime). There is also no way to make them, let alone show that there are limitless numbers of them. But Euclid's proof is extraordinarily simple, almost stupidly so. Assume, he argued, that there are only a finite number of primes: $P_1, P_2, P_3 \ldots P_N$. Form a number that is their product; then add 1 to it and call that number $Q = [P_1 \times P_2 \times P_3 \times \ldots \times P_N] + 1$. That number is not divisible by any of the prime numbers listed (there would always be a remainder of 1). But if Q isn't prime itself, it must be divisible by another prime—one not listed. This contradicts our assumption, so our assumption is impossible; thus, there must be an infinite number of primes.

What Euclid did in his proof was to construct a new number with properties that force us to recognize a fact about *all* numbers, creating a single case that shatters an abstract assumption. It is a form of proof that has been essential in the history of mathematics whenever principles of extraordinary abstraction were examined. These proofs create a universe ("assume that these elements are finite") and then play with its elements, often generating surprising and unexpected results. The play seems, at first, quite simple, even mundane. Then the gestures ("assume that ..." and "form a number ...") act like lightning bolts illuminating the terrain, startling us and starkly revealing objects we had overlooked. The result can be profound. The creation of a new number, Q, is something unpredictable, a gesture that uses the laws of the number game to astonish us.

This proof is not produced according to formula. In fact, there are dramatically different ways to prove the same result that involve constructions of still more unusual numbers. For example, there are numbers called Fermat numbers, which are of the form $2^{2^n} + 1$. For $n = 1$, the Fermat number would be 5 $(2^2 + 1)$, for $n = 2$ such a number would be 17 $(2^{2^2} + 1)$, for $n = 3$ the Fermat number would be 257 $(2^{2^3} + 1)$. This series grows extremely fast and has many unusual properties, one of which is that no two numbers in the series share any divisors. Thus, we can find no number other than 1 that will divide both 5 and 17, or both 17 and 257, or 17 and 65,537, or 257 and 65,537. This means that for every Fermat number there is a prime number (other than 2) that divides it but does not divide any other Fermat number (otherwise they would share a divisor). So for any list of n Fermat numbers, there are at least n distinct primes, and since there are an infinite number of Fermat numbers, there are an infinite number of primes.

This proof, by the mathematician George Polya, is not as satisfying as Euclid's: it involves a series of numbers that at first seems irrelevant. It is more "tricky," its moves less motivated by a simple consideration of how we count (by adding, as in

Euclid's proof, 1 to a previous number) than by how we explore properties of numbers. But it is more satisfying in another respect: we have a concrete series of numbers, one that we can calculate, that goes on forever, and we have connected the number of primes directly to that concrete series.

There are different styles to these two proofs, different emphases, even different presentations of the infinite: in the first we never see an infinite series of numbers, while in the second we do. The first is more fundamental, the second more suggestive. The first does not really lead anywhere; it does not cause us to ask any more questions. The second does: it intrigues us about this other series of numbers with this unusual property. Neither of these proofs, though, was mechanistically "produced" in any way. Their styles and approaches, moreover, have much to do with the styles of "doing" mathematics in their very different times, and with their places in larger systems their creators envisioned. Euclid's proof is about building numbers, Polya's about connecting them.

These elementary examples emerge out of the world of arithmetic and number and are somewhat unassuming; the result of their inventiveness, after all, is not particularly revelatory. If the primes ever came to an end, it would be far more startling: think of all those infinite numbers higher than the last prime; how could only the primes listed be used to compose any number, no matter how large? (That is the motivation behind Euclid's proof.) Grand proclamations of an infinite number of primes are less intriguing than the styles of the proofs themselves. They suggest ways of thinking about numbers, offering approaches to the ordinary that become extraordinary.

Understanding these simple examples means beginning to see mathematics as part of what is commonly called the humanities. This already brings it closer to music. Far from being a mechanistic system, wound up and ready to click along, mathematics raises questions of style and taste. A mathematical proof does not stand in isolation; it answers one question and points to others. It takes certain things for granted while subjecting others to scrupulous examination. It is a judgment about importance. This is one reason that many mathematicians find computer-generated proofs of various facts about the number system so unsatisfying; simply carrying out calculations does not reveal anything significant about that system; it just does the job. These "brute force" exercises, as they are often called, stop short of carrying out the mathematician's vocation, which is to present an ordered view of the universe being explored. Such proofs may solve problems, but they are considered by many mathematicians to be the equivalent of technical exercises, asserting a truth or method without making a significant contribution to our understanding of the mathematical universe.

Differences in style like those in these proofs dealing with primes are not minor matters. They affect the substance of mathematical work; they are related to the shape of the mathematical universe. Consider again the example of geometry, which was once just the study of space and line familiar from experience. We know that non-Euclidean geometries revealed that there is nothing "real" about that mathematical universe. Euclidean geometry was manufactured, a creation of assumptions and rules, an idealized image that was shown to be just one possible universe among others. And, of course, we now know too that our physical world really is non-Euclidean; no space scientist could survive using purely Euclidean assumptions. These differences affect the way arguments are made and proofs constructed.

It might be helpful to imagine a mathematical universe as being a context or a setting within which intellectual exploration or play takes place. A mathematical universe, which may consist of a particular axiomatic system and the deeper, unstated assumptions underlying it, provides the mathematician with information about what is permitted, what is forbidden, what is "known," and what is "unknown." Our premises and propositions depend upon our inheritance or choice of such a context. In the context established by classical Euclidean geometry, the concept of a four-dimensional sphere makes no sense—it doesn't fit what might be called the grammar of a system which is designed to deal only with two-dimensional space and which has no way of creating or studying other dimensions.

Because context is so important, aspects of mathematical truth may alter over time. When ideas about possible geometries changed, so did the immutable facts of their Euclidean ancestor. Every mathematical truth may be subject to boundaries, spoken and unspoken, acknowledged or even unknown. The history of mathematics is relevant to our understanding of mathematics. Although there are some truths and results that apparently stand outside time, whatever the context, the posing of questions, the methods of exploration, even the accepted methods of proof are just as bound up with a mathematician's time and place as a musical style is with a composer's.

Paradoxically, this is one reason why so few mathematicians ever study the history of their own discipline. The apparent uniformity of truth through time might seem to suggest that a geometer might as profitably study Descartes as Lobachevsky, or a number theoretician might as usefully read Euler as Hardy. In fact, the language of mathematics today bears so little resemblance in style and form to the languages of its past that it would take a great deal of effort to "translate" the mathematics of the past into contemporary terms; doing so would also be of very little use. Earlier notions of style also often masked mistaken assumptions or distorted perceptions. Today's preferred forms of presentation doubtless hide errors as well. Much twentieth-century emphasis, for example, is not on practical application but on abstraction, form, and the creation of structures that allow disparate connections to be made. We have (even in this book) a structural bias.

One example of shifting context and the transformation of

mathematical styles was discussed by the mathematicians Philip J. Davis and Reuben Hersh in *The Mathematical Experience*. They present a simple theorem of arithmetic that has been generally known as the Chinese remainder theorem. First formulated in the *Sun Tzu Suan-ching*, a Chinese volume written about 2,000 years ago, it provides a method for finding a hidden number by using the remainders produced when the number is divided by other numbers. (For example, if I am thinking of a number less than 60, and I tell you the remainders received when the number is divided by 3, 4, and 5, then you can figure out the number itself.) The theorem is about calculation; it uses numerical patterns without making them explicit.

Davis and Hersh go on to give several other examples of the theorem in the West. The thirteenth-century Italian mathematician Leonardo Pisano (more commonly known as Fibonacci) retained the Chinese concreteness (though his vocabulary bears today a confusingly antique aura):

> Let a contrived number be divided by 3, also by 5, also by 7; and ask each time what remains from each division. For each unity that remains from the division by 3, retain 70 [i.e., multiply the remainder by 70]; for each unity that remains from the division by 5, retain 21; and for each unity that remains from the division by 7, retain 15. And as much as the number surpasses 105, subtract from it 105; and what remains to you is the contrived number.

Fibonacci suggested a "pleasant game" of guessing secret numbers from such calculations. But he placed his theorem and his game in no context other than mere calculation.

In the early eighteenth century, the great mathematician Leonhard Euler put the same result in more general abstract form. It is closer to modern mathematics in its insistence on recognizing the abstract pattern in these seemingly unrelated figures. This example, along with the others provided by Davis and Hersh, need not be understood here, just recognized as representing different mathematical universes in style and approach; in fact, their opacity and difficulty are part of the point:

A number is to be found that, when divided by a, b, c, d, and e, which numbers I suppose to be relatively prime [they share no common divisors], leaves respectively the remainders p, q, r, s, and t. For this problem the following numbers satisfy:

$$Ap + Bq + Cr + Ds + Et = m \times abcde$$

in which A is a number that divided by $bcde$ has no remainder, by a, however, has the remainder 1; B is a number that divided by $acde$ has no remainder, by b, however, has the remainder 1 ... which numbers can consequently be found by the rule given for two divisors.

This is still close to the earlier formulation, though it requires more exegesis. But a general pattern rather than a specific case is being emphasized. The next two examples, I won't attempt to explain (even Davis and Hersh strain at them). First one by R. E. Prather:

If $n = p_1^{\alpha_1} p_2^{\alpha_2} \ldots p_r^{\alpha_r}$ is the decomposition of the integer n into distinct prime powers: $p_i^{\alpha_i} = q_i$, then the cyclic group Z_n has the product representation $Z_n \cong Z_{q_1} \times Z_{q_2} \times \ldots Z_{q_r}$.

Then one by another contemporary mathematician, E. Weiss:

Of $S = \{P_1, \ldots P_r\}$ is any finite subset of $\pounds$, then for any elements $a_1, \ldots a_r \in F$ and any integers $m_1, m_2, \ldots m_r$; there exists an element $a \in F$ such that

$$v_{P_i}(a - a_i) \geqslant m_i \qquad i = 1, \ldots r$$

$$v_{P_i}(a) \geqslant 0 \qquad P \notin S, P \in \mathcal{L}.$$

These last two formulations not only discard all practical techniques for finding what "exists" but alter the theorem so its emphasis is not on calculation or on the establishment of certain relationships between numbers but on various highly abstract systems of cyclic groups and fields and other mathematical constructs. There is, as Davis and Hersh suggest, an element of structural and abstract "zealotry" in these sorts of formulations which one does not need to understand to sense. What is involved here are not just differing styles of mathematics but different categories of judgment, different attitudes to fundamental facts. The individual theorem actually fades in significance in the recent formulations; its importance is as part of a system of mathematical structures rather than as an indication of numerical truth.

Mathematical style is far more important than it usually seems. It is intimately connected to the essence of mathematical work. It defines conditions and expectations. It presents a set of rules, of course, but it also does something more: it reflects what is considered important at a particular historical moment and shapes the evolution of future inquiries. It resembles, in this way, musical style.

OF COURSE, mathematics is not just a matter of style. There is, as we all know, progress in mathematics, an increase in knowledge over time. This is obviously a quality mathematics does not share with music—at least not literally. But we need to understand how that knowledge is reached, how questions are formulated, and how answers are analyzed. For one similarity between math and music is in how themes are developed and their implications explored.

At the very beginning of this inquiry I raised questions about how we compare objects and define their similarity, how comparison requires a process of abstraction and connection—creating what mathematicians call a mapping from one object to another. A mapping is simply a way of connecting two distinct things: we might map the letters of the alphabet into the set of integers, giving each letter a number; this mapping could be used as a kind of code. We might map the curved lines on recording grooves into waveforms in the air; this mapping would be accompanied by intermediate mappings of the physical curves into electrical signals. We might map a sphere onto a section of the plane; this mapping could be a literal map, which distorts the globe onto a flat piece of paper. We could even map the various mappings we make of the earth onto the various names we have given them, such as the Mercator. Mapping is an essential technique of mathematics. In every case of a mapping, we have first to define certain characteristics of the sets we are mapping and determine the internal relationship between their elements; only then can we proceed to make external mappings.

Suppose, for example, we are given two surfaces—a torus (a doughnut shape) and a sphere. Are they fundamentally the same or different? Do they share properties? Are the properties they do not share of any importance? We think we understand these questions well. Both are "smooth," but one has a hole, the other doesn't. But what does this really mean? We know that a string can be put through the hole, wrapped around the doughnut and a knot made, and the string cannot be pulled off without ripping the doughnut apart. On the surface of a sphere any tied string will simply slide off. But how can we describe this difference mathematically? How can we also find ways of differentiating and understanding many objects—a double doughnut, a string tied in knots? The answers have something to do with the kinds of mappings that are possible between spaces and how such mappings make connections palpable.

There is no easy way into this problem. Have we understood the differences clearly? What is a surface? What is a knot? What is a hole? Can we discuss these things without recourse to sensory experience, without saying, "Well, you know, you can poke something through a doughnut hole, so obviously it's different from an unbroken surface"? Is there a way to translate our understandings into a precise vocabulary—to map our intuition into knowledge? How do we specify that a surface is smooth or a line unbroken? These initial challenges require moving from the concrete world to the world of mathematical concepts.

We do this by attempting to find what is essential: we compare disparate objects that seem to share a certain characteristic, say, holes. We attempt, gradually, to understand what a hole is. In music, as we shall see, we do the same thing with sounds. We listen to the sounds of overtones and chords and begin gradually to find similarities in tensions and relaxations. Out of the world of simple sensation, we discern connections and then create a form of knowledge from what was once just a collection of impressions. A musical system develops out of abstracted similarities and patterns. So does a musical composition.

But let us focus more closely on the process in mathematics. The example of surfaces is useful in trying to understand mathematical thinking because it seems, at first, so obvious, so clear to simple intuition. At stake is nothing less than our intuition of space, what might be called our sense of smoothness.

This sense has actually been a major obsession of mathematicians throughout history; smoothness and regularity of space and time are associated with a notion of continuity. The issue itself began with the Greeks; Zeno posed the most famous paradoxes of continuity, which we know about through their description in Aristotle's *Physics*. Imagine a race, Zeno suggested, between the swift Achilles and the slow tortoise, who has, in all fairness, been given a head start. Achilles could presumably still win the race, but Zeno argued this is impossible. He can never overtake the tortoise. For by the time he reaches the spot from which the tortoise began, the tortoise will have moved forward to another position. By the time Achilles reaches that spot, the tortoise will be at another. Achilles is doomed to be always one location behind the tortoise. The paradox is similar to another Zeno proposed: before traversing a distance—a mile, say—an object must pass the half-mile point. Before that it must pass the quarter-mile point, before that the eighth, and so on. In fact, any motion requires a traversal of an endless number of points in a finite time. Thus, it may reasonably be concluded, all motion is impossible.

These paradoxes succeed in baffling us because they present an idea of space and time that is utterly at odds with our everyday experience. They proclaim an endless succession of detached points instead of a seamless, continuous motion, in which there is no such thing as a "next instant" in time or a "next point" in space.

Our experience of numbers, though, is quite different, and this difference has had important implications for the development of mathematics. Integers are discrete. We can always count the next one; there are spaces between them. In *Number: The Language of Science*, the mathematician Tobias Dantzig made an elegant musical distinction between the geometrical intuition of continuity and the logic of arithmetic, which seems so peculiarly full of gaps. "The harmony of the universe knows only one musical form—the legato; while the symphony of number knows only its opposite—the staccato," he wrote. "All attempts to reconcile this discrepancy are based on the hope that an accelerated staccato may appear to our senses as a legato." When faced with the staccato sequence of numbers, we have an impulse to fill in the gaps so that numbers can begin to resemble our experience of space, so powerful is our quest for smoothness and continuity.

We have already seen how the rational numbers are formed from ratios of integers. Numbers like $\frac{2}{3}$ and $\frac{34,234}{38,792}$ also seem to fill in the space between integers (in this case, between 0 and 1). In fact, once we combine rational numbers and integers, we do seem to have continuity and smoothness: it is no longer possible to say which number is next, and there no longer seems to be any "space" around a number. What number comes after $\frac{2}{3}$? No matter what number we name, we can always name a number closer; $\frac{2,001}{3,000}$ is very close, but $\frac{20,001}{30,000}$ is still closer. In addition, no matter how small a "distance" around a number we might specify—say within .00001 of $\frac{2}{3}$—we can always find a number closer than that distance; we can even find an infinite number of numbers within that distance. This might make the number system seem quite smooth indeed, each region around a numerical point packed with an infinite number of neighbors. It certainly fits our intuition of spatial and temporal smoothness, our sense that it is possible to slide over a surface without interruption and move infinitely close to whatever point we like.

But irrational numbers disrupt our intuition. Such numbers seem to upset this numerical universe, to show that this apparently packed and smooth space between numbers is actually full of peculiar gaps. Moreover, there seems to be no clear way to construct or reach these irrational numbers from the rationals, the way the complete world of rationals could be constructed from the integers. The square root of 2 cannot be created with the ratio of two integers; it is actually called incommensurable— unmeasurable with rational units of space. The incommensurables seemed to lay beyond what had been regarded as appropriate mathematical language; they disrupted an order that was both pleasing and rational. The Pythagorean belief in the primacy of whole numbers and their ratios had been shattered by a sort of mathematical dissonance that could not be absorbed or interpreted. This dissonance made the harmonies of the rationals more pleasing, but it also suggested that the irrationals had to be understood, connected somehow to the widening world of mathematical knowledge.

One might represent irrational numbers by geometric pictures—a square with sides of length 1 has a diagonal with the length of the square root of 2, for example—but how could they be represented using the rational numbers, using the processes of ratio and proportion upon which the number system seemed to be based? An irrational number could be specified with an equation or a diagram, but where did it fit in this apparently smooth universe? How could the preexisting system be stretched in a way that made sense, the way a dissonance in music can be incorporated into a composition without creating a sonic shipwreck? We know an extreme dissonance like the tritone was, in fact, not easily accepted into the musical system except as a representation of violence (indeed, it was not until the twentieth century that it became commonplace and almost lost its power to shock); the irrationals posed a similar challenge.

For all the perplexity they caused among the Greeks, attempts were made to close the gaps in the number system out of which the irrationals seemed to emerge. For example, Eudoxus tried to "approach" the irrational, to reach it from the rational numbers. He created a ladder of whole numbers like this:

1	1
2	3
5	7
12	17
29	41

and so on.

The left hand of each rung is the sum of the preceding rung $(1 + 1 = 2, 2 + 3 = 5$, and so on). The right hand of each rung is the sum of the left rung with the preceding left rung $(1 + 2 = 3, 2 + 5 = 7$, and so on). But it is the ratio between the two numbers on a rung that is of interest. As one moves down this ladder, they proceed: $\frac{1}{1}$, $\frac{3}{2}$, $\frac{7}{5}$, $\frac{17}{12}$, $\frac{41}{29}$, and so on. One ratio is too high for the goal, and the next too low, but this series gradually gets closer and closer to an irrational number—the square root of 2—without ever reaching it. This is a clever way of characterizing an irrational, discerning an esoteric pattern that points to it; uttering the unutterable by descending to it, step by step. But Eudoxus's ladder is, in fact, the *only* way the irrationals can be reached through the rationals—a process, like that created by the hapless Achilles in the Zeno paradox, moving steadily closer, never quite touching the object of rational desire. There is a bitter truth for idealists in irrational numbers: they can only be written as infinite decimals with no repeated pattern. For example, there is neither pattern nor end to the decimal for π: 3.14159265358 . . . The irrational decimal, lacking a pattern, seeming almost random in its approach, never ends because it can never completely attain its goal. The irrational lies always just out of reach.

The notion of an unreachable limit, of course, is at the heart of Zeno's paradoxes, though it seems there somehow misapplied to space when it should just be applied to number. But the problems of the continuity of space and the continuity of numbers are related; we measure space with numbers and imagine numbers spread through space. In fact, the number line is probably something we have all come across: a line with notches for each integer extending infinitely in each direction, in which every number can be represented by a single point.

Continuity, in both space and number, is a matter of coherence; it poses the challenge of matching mathematical thinking to everyday experience. Not until the seventeenth century did the issue begin to make more sense, in the worlds of both space and number, in the calculus invented simultaneously by Newton and Leibniz. While we use Leibniz's notation and even his conception as a method of calculation, we are partial to Newton's interpretation as an attempt to understand motion. Newton called changing variables such as space and time fluents—for everything, he believed, is truly fluid in motion. Instantaneous rates of change, such as instantaneous velocity, he called fluxions—we now call them derivatives.

Calculus is a way of grappling, in part, with the issues raised by Zeno. When we drive a car and glance at the speedometer reading 55 miles per hour, what does that mean? Speed is a measurement of the amount of space that can be traveled in a certain time. We calculate speed by taking the distance traveled between one moment and another and dividing it by the amount of time that has gone by. Hence we might name a speed one meter per second, by which we would mean that in the passing of a second the car travels one meter. But when we glance at the speedometer, we are not really dealing with the passing of time, or with the passing of distance; we have nothing more than the instant we look at the dial. At that instant, the car doesn't pass through *any* space: we can find a precise point at which the car is located at the instant we check the dial. Zeno said that if we can specify precisely where an arrow is at a particular instant, it has to be at rest at that instant; hence it has to be at rest throughout its flight. Why should cars be any different? What does it mean to say a car is traveling at a certain rate of speed when, in theory, no time is passing and no distance is traveled?

That is one sort of problem calculus addresses: it allows us to discuss instantaneous events. But the methods for dealing with the instantaneous were, even at the time calculus was invented, not fully understood. At least in Newton's profound formulation, they depended on a conceit: let us pretend, he argued, that this instant of no duration is actually an interval of *very small* duration, so small we can call it infinitesimal. The infinitesimal is smaller than any number that we can name but is greater than zero. Newton went on to say that when it suits our purposes we should ignore the infinitesimal completely and set it equal to zero. This conceit turns out to be extremely useful and at first quite surprising. An instantaneous velocity is simply defined as the infinitesimal distance traveled in an infinitesimal time; what seems meaningless ($0 = \%$ is what this arithmetic of infinitesimals implies) becomes important when we deal with the infinitely small but not nonexistent. It gives answers that make sense.

The idea of the infinitesimal is an intoxicating "trick": it involves pretense on all sides. But it does seem to resolve something about continuity: gaps can be filled with infinitesimals. And so it was not examined too closely. Leibniz tried to explain the "infinitely little" to Queen Sophia Charlotte of Prussia; she said that on that subject she needed no instruction—the behavior of courtiers had made her intimately familiar with it. That is about the level on which early mathematical manipulation of infinitesimals took place; these odd mathematical creations were used as placeholders, minor functionaries, little more than courtiers in the court of calculation.

But so intoxicating was the concept that it was asserted that nearly anything could be constructed using the infinitesimal. The first calculus book, written in 1696 by the marquis de L'Hôpital, defined a curve as "the totality of an infinity of straight segments, each infinitely small" (a formulation which echoed Galileo's). The infinitesimal also inspired some backlash. One critic of the period attacked "mathematicians who, in their kingdom of abstraction, assume indivisible substances which are without parts, without length, and without width." The most celebrated critic of cavalier play with infinitesimals was George Berkeley, who called them "the ghosts of departed quantities" and wrote, "He who can digest a second or third fluxion ... need not, methinks, be squeamish about any point in Divinity."

Strange things took place when the infinitesimal was treated cavalierly. Because this object was treated like any other number until it became convenient to set it equal to zero, the issues involved in continuity were never addressed. What would it mean to come infinitesimally close to something—say within an "infinitesimal" distance? What would happen when an infinite number of infinitesimal objects were added? When did they reach a sum and when didn't they? We know, for example, that if we form the sum $1 + 2 + 3 + 4 \ldots$ we never reach a conclusion but we do reach something that gets ever larger. We also know that if we add $\frac{1}{2} + \frac{1}{4} + \frac{1}{8} \ldots$ we come closer and closer to 1. But *how* close does the latter sum come to 1? Can we say it is equal to 1? Does it differ by an infinitesimal— or by nothing?

These questions grew out of an inadequate understanding of infinitesimals and the infinite. In 1703, one Italian mathematician used the known algebraic expression

$$1/(1 + x) = 1 - x + x^2 - x^3 \ldots$$

(This might look baffling, but multiply the infinite right side by the fraction's denominator and watch all terms cancel except 1.)

He simply set $x = 1$ and came up with the absurd statement:

$$\frac{1}{2} = 1 - 1 + 1 - 1 \ldots$$

What was to be made of this?

Leibniz asserted there was nothing peculiar about it. He reordered its terms, obtaining:

$$(1 - 1) + (1 - 1) \ldots = \frac{1}{2},$$

which would mean that

$$0 + 0 + 0 + 0 \ldots = \frac{1}{2}$$

Leibniz's conclusion was not that there was something fundamentally wrong with the conception of infinite series reaching a particular sum but that this summing of zeros was a profound result. First, he said, summing an infinite number of zeros and reaching $\frac{1}{2}$ proves that the world could be formed from nothing. Second, $\frac{1}{2}$ is the result because it is an "average" between the sums of 1 and 0, between which this infinite sum seems to vacillate.

Euler was horrified by such claims, but he claimed similarly absurd results, such as $\frac{1}{4} = 1 - 2 + 3 - 4 \ldots$ How, then, could anything be clearly understood about the irrational numbers, which required such infinite calculation to be "reached" and specified? What were we to do about π or the square root of 2 in this mess?

This historical and philosophical sketch of the problem leaves out many nuances and issues. But a fundamental issue was, as Galileo put it, the "building up of continuous quantities out of indivisible quantities," the creation of continuity—of space, of the world of our experience—out of Berkeley's "ghosts," those imaginary infinitesimals endlessly combined. In retrospect, it is clear just how much an artifice this mathematics was, a construct that was attempting to model the world by remaking it.

At the time, what was pretended—and what I would guess is always pretended in mathematics—was that a particular set of suppositions along with a vocabulary or grammar is more than just a style but is regarded as a revelation of the truth. It was as if a practice were taken as law, an invention of the intellect as a manifestation of nature itself. This is always a risk in mathematics: conceptual objects like infinitesimals become reified. Abstractions created for particular reasons—as tools, intermediaries, conveniences—begin to take on a life of their own. The abstract becomes concrete. Mathematics proceeds on the basis of hypotheses with the confidence that each step is as firmly placed as the last. Often the resulting path is, at the very least, full of brambles and thorns of little use to later generations.

Thus, mathematics is as subject to upset and revision as are theories of science and styles of art. This is a surprising fact if we are used to thinking of math as a purely linear discipline confidently moving through—and outside of—history. For in math, as in music, a style can become established as law—we have seen examples of such styles in our discussions of prime numbers, the Chinese remainder theorem, and the evolution of the infinitesimal—a style can become so dominant it is treated as "natural" and cannot be violated without stepping over the boundaries of the field itself. Indeed, style can be treated as if it were a natural phenomenon—as rooted and invariant, say, as tonal musical language has been regarded, as if it were the only possible way of organizing the natural priorities established by the vibrating string's overtone series. The difference is that in mathematics there is an arbiter that, in music, we may only see metaphorically at the end of our journey: the truth.

Truth in mathematics, though, is often quite different from truth in physics. There are, of course, "facts" that come to light in mathematics—figuring out, for example, that a number is prime, or knowing (as no one actually does) whether every even number but 2 is the sum of two primes. But these sorts of truths are accompanied by concepts like the infinitesimal and elaborate strategies like the calculus, which take a certain shape and direction because of matters of taste and practicality and style. These matters are so entwined with issues of truth that it is all but impossible to separate them; decisions are made for combinations of reasons.

Examining the inner life of mathematics requires attending to these different aspects of truth: what we are given and what we make of what we are given. (We will find this is the case in music as well.) The saga of the infinite and the infinitesimal, the continuous and the discrete, continues through a variety of variations on this theme.

The limitations of the procedures and the interpretive concepts, for example, were becoming clear by the nineteenth century, largely because of a deeper understanding of what the terms "infinitely small" and "infinitely large" actually meant. An instantaneous event, the sum of an infinite series, the precisely calculated irrational number—these gradually came to be treated not as concrete objects that would be grasped and manipulated but as the results of *processes*, goals for a sort of mathematical yearning. Like the figures on Keats's Grecian urn, an infinite series would strain, ever reaching for, never attaining its goal. The idea of the infinitesimal evolved into a dynamic notion of limit and approach, in which mathematicians agreed that infinitesimals would never simply be set equal to zero and infinite series would never be treated as ordinary finite sums.

So again: what does it mean to come infinitely close to something? The old question has lost none of its vexing power. During the nineteenth century, Augustin-Louis Cauchy and Karl W. T. Weierstrass answered this question in so compelling a fashion that they created a style of mathematics that is still dominant today. If we want to know about instantaneous velocity, for example—that instant we glance at the speedometer— we must gradually allow the time interval and the distance traveled to *approach* zero; as a smaller segment of each is taken, we watch what happens. This sequence of measurements has a limit; it seems to approach a number without necessarily reaching it. That number—the limit of this sequence—is the derivative, which Newton called the fluxion. In this interpretation, we should never "set" a number equal to zero but instead simply examine the process. This is still a conceptual invention—imagine the absurd physical experiment that would be needed to confirm the speedometer's reading—but it is a profound invention. This is a shift in focus from what could be called a "classical" approach, dealing with objects that possess immutable properties and proportions, to a sort of "romantic" approach, in which development and transformation become central. (A similar transformation occurred in aesthetic taste and structural logic in music—but that is getting ahead of ourselves.)

This shift of attention to process deserves closer examination. The assertion that an infinite process or series reaches an end, or limit, if it can be made "arbitrarily" close to that value is a strange one, particularly since we are dealing with very large processes to describe instantaneous facts. But the definition Weierstrass created was actually quite simple, because it put abstract, esoteric concepts into concrete, arithmetic terms.

Technically this definition appears in a contemporary text as follows: "We say that

$$\lim_{n \to \infty} S_n = L$$

if for each positive ε there is some integer N depending on ε such that $|S_n - L| < \varepsilon$ whenever $N \leqslant n$." Such a mathematical hieroglyph may appear indecipherable, but it is a rigorous restatement of the conditions just described. It is simply a way of saying that when a series of numbers S_n has a limit of 0, as n gets larger and larger, for any number we choose (ε), no matter how small, we can find a point in the series beyond which all members of the series will be within that small distance of the limit. If, for example, we consider the series of numbers 1, $\frac{1}{2}$, $\frac{1}{3}$, $\frac{1}{4}$, $\frac{1}{5}$..., we can assert that this series of numbers of the form $1/n$ where n is an integer is never equal to 0, but it approaches 0 as n gets larger and larger. The crucial fact is that we can make it as close to 0 as we like: for any minute distance we may specify—such as .0000000001—we can find a number far enough down the series so that this distance from 0 is reached and passed. Similarly, we may look at the series of numbers 1, 1.4, 1.41, 1.414, 1.4142, 1.41421 . . . as approaching the square root of 2—the irrational number mentioned earlier.

In fact, any irrational can be expressed by infinite rational sequences reaching a limit. Nothing is ever set equal to zero; our intuition is appealed to rather than discarded. The limit is a simple matter of calculation. This is also what we intuitively understand when something gets closer and closer to a goal. Give me a distance, it challenges, which you think is closer than I will ever get, and I will show that the numbers get even closer than that. This method involves no infinitesimals or other strange objects, only Greek letters like ε and δ, which symbolize the challenge. The definition actually succeeds in specifying what role the infinitely small and infinitely large play in our number system, and it makes rigorous the notion of approach. It also makes the practice of mathematics more self-conscious. The definition of a limit asserts that we can *find*, or we can *make*, or we can *specify* a point beyond which the series lies within a certain distance of its limit. Once this notion of limit is defined, it becomes possible to view irrational numbers in a new way. The irrational is not something outside the rational universe; it just must be *calculated* differently. If we map numbers into points on a line, an irrational is just another point, specified by using the sorts of calculations Weierstrass made available.

But there are other complications. Adding the irrationals is not just a matter of adding a set of numbers to our conception of the number system: it means adding more numbers than any that have yet been included. There are more irrationals than rationals, more irrationals than integers. By adding the irrationals, we close up the number line, rendering it both smooth and complete, thereby transforming something fundamental in its structure.

This is a peculiar assertion. But measuring infinitely large and infinitely small sets is a peculiar activity. One might think, for example, that there are fewer even integers than there are integers. If we look at the integers:

$$1, 2, 3, 4, 5, 6, 7 \ldots$$

and then look at the even numbers:

$$2, 4, 6, 8, 10, 12 \ldots$$

it is easy enough to see that the set of even numbers is completely contained in the set of all numbers. Ordinarily when one collection contains another the latter has fewer elements.

But it turns out, as every high school student learns, that there are exactly the same number of integers in each set: there are the same number of even integers as there are of even and odd integers together. The reason: for any integer n, we can match it to an even number $2n$; and for any even number m, we can match it to an integer $m/2$. This "one-to-one correspondence" is how we know the sets have the same number of elements; it is also how we count objects at all.

We might surmise, then, that in infinite sets the whole is not necessarily greater than its parts. In fact, this is one proposed definition of a set that is not finite: an infinite set can be put into a one-to-one correspondence with one of its subsets. But our intuition rebels. We discussed earlier that while one can always name a "next" integer, one cannot always name a "next" rational number: rationals seem more densely packed, as if there were infinitely more of them. But, surprisingly, this is not the case: there are precisely the same number of rational numbers as integers. And the brilliant nineteenth-century mathematician Georg Cantor, who took the properties of the infinite as his domain, proved it so. Cantor cleverly placed rational numbers in such an order that one could name a successor and a predecessor to every rational number and "count" it with an integer. His arrangement deliberately avoided all notions of size, since by size there can be no listing.

Cantor established a square in which each rational appeared:

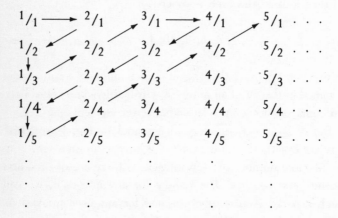

By moving along this infinite square following the arrows, Cantor could list every rational number. And he could "count" them, that is, he could match each rational with an integer and vice versa. Thus, he could conclude there are the same number of rationals as integers.

But Cantor also went on to prove that there is another *kind* of infinity, one larger than these; that infinity is the number of *real* numbers, including those irrational numbers that so plagued the Greeks and cannot be represented as fractions of integers. In this greater infinity, one can't find the next number no matter *how* one attempts to order the numbers; the density of real numbers is, in fact, the smoothness of the line. Its infinity of points being nondenumerable, it cannot be numbered, or counted, or listed.

Cantor's classic proof of the existence of this greater infinity assumes first that the conclusion is incorrect, that the number of real numbers has been "ordered" like the rational numbers and integers. We can write any of these real numbers as infinite decimals, such as π, 3.14159265358 . . . So, we assume, we can create a list of every number: n_1, n_2, n_3, and so on, each written out in decimal form (with a minor restriction that need not concern us here). But as with Euclid's proof about primes, we create a new number that contradicts our assumption. That new number is nowhere on this list.

Here is how. The new number will differ from the first number on the list in its first digit; it will differ from the second number on the list in its second digit, from the third in its third digit, and from the nth in its nth digit. By going down this "diagonal," a number is created that is different from *every* single number on the list by at least one digit. Therefore, it is not included on our list, which contradicts our assumption that we could list and count all such numbers. Hence, we cannot count the real numbers: there are "more" of them, they are uncountable.

Cantor's proof is simple and brilliant. It has a brusque clarity to it, proving something astonishing by means of something seemingly mechanical. Cantor had a gift for turning remarkable assertions about intricate sets into almost elementary diagrams. At first the proof is also dissatisfying for these reasons, for there is nothing in the way this number is chosen that gets at the essence of what is being shown. Euclid constructed a new number in much the same way as all numbers are constructed—by multiplying primes. Cantor's new number is important only because it is new; it has no other, larger significance; it is created using a "trick" and seems to have, despite itself and as if by accident, revealed, as it were, a new continent.

Nevertheless, the proof attacks the heart of the issue— ordering and listing—and does so with exceptional elegance. The very notion of ordering and listing means inherently that we can always find a "next one" and that we somehow know how to proceed from one to the other. It implies a grounding in some point of origin and then a rule of some kind that lets us move forward. This notion is fundamental to mathematics and its methods. So to prove, as Cantor did, that there can be no such listing becomes quite profound. The result of his proof opens the door to an astonishing universe in which there is no way to move in an orderly fashion from one object to another. Cantor dramatically changed the way we think about infinities. And he did it by changing the way we think about something as mundane as a list. The entire proof recalls, in a distant way, Zeno's paradoxes, which depend so heavily on specification of "next points" and "next instants" in a world where there is no next.

This proof, in its elementary arithmetic fashion, also uncovered a completely new number: the number of irrationals. This number is greater than the number of integers, though both are disconcertingly infinite. At the heart of this discovery of a new number is a refinement of the concept of continuity—in which space or time itself is so dense with points that any area surrounding a given point contains an infinite number of other points, which, in turn, connect "smoothly" with one another. If any points "approach" another point in an infinite sequence, that limit point is also included in the continuous collection of numbers. Anaxagoras anticipated Cantor's discovery when he wrote, "Among the small there is no smallest, but always something smaller. For what is cannot cease to be no matter how far it is being subdivided." Once we include the irrationals we have a system of numbers that can properly be called a continuum; the number line now is without gaps, without twists; it is seamless. Of course, the irrational numbers still remain peculiar. While modern notions of limit and continuity satisfy our rational minds, they prove inadequate as an aid to understanding the strange qualities of the number line, which seems to divide into different types of numbers: those we define through limits and infinite series, and those we define using simple ratios.

It was not until the end of the nineteenth century that there was a satisfactory definition of number that treated all points on the number line equivalently. The mathematician Richard Dedekind actually used the notion of continuity to define irrational numbers. In 1872, in his essay "Continuity and Irrational Numbers," Dedekind wrote that "the comparison of the domain of rational numbers with a straight line has led to the recognition of the existence of gaps, of a certain incompleteness or discontinuity in the former; while we ascribe to the straight line completeness, absence of gaps, or continuity. Wherein then does this continuity consist?"

Splitting a line into two portions at any point, he argued, divides the set of points into two parts, one of all points to the left of that point, another of all points to the right. This is also, he noted, a characteristic of numbers; whenever a number is specified—say, 2—there are two other collections of numbers created. Members of one are less than 2 and less than every number in the other class; the others are greater. The number splits the numerical universe. Dedekind decided to define that split as the number itself. This avoided all the knotty philosophical and mathematical problems involved with the standard conception of number. Each number is a "Dedekind cut."

Dedekind's insight used both the notion of limit and the sense that the line and the set of all numbers form a continuum. There was no distinction between the rational and the irrational; both were cuts, both divided the line into two parts. Dedekind's definition transformed our conception of numbers, enabling us to identify each number with points on a line, creating a series of relations—spatial and arithmetic—with every other. It was as if Zeno's paradox of the arrow moving was, with Dedekind's help, taken literally: every point through which the arrow passes divides its trajectory into a before and an after. There is no "immediately preceding" or "immediately following," for we have effectively frozen the path for study the moment we define the arrow's position. This has become a preferred theoretical definition of numbers, upon which the modern system of arithmetic has been built. Dedekind turned the bizarre into the ordinary. Now irrationals too could be considered as just other points on the continuum. Dedekind's methods homogenized the number system, at least for mathematical inquiry, making each point on the number line open to equivalent consideration. There was no center, only a smooth array.

WE BEGAN our inquiry by asking questions about space and ended with the extraordinary efforts to conceive of a system of numbers that can be thought of spatially—as continuous rather than discrete; the number system was made rational. I do not claim that these issues were the motivating ones in mathematical work, and the history of mathematics is far more complex than I have suggested in this brief sketch. But it should be obvious—as a mathematician might say—that mathematical thinking bears an intricate but indirect relation to the everyday world, that it involves a large element of innovation and risk and invention and faith, that mathematical style is based upon an implicit conception of the mathematical universe, that a particular inquiry in mathematics is the result of intense concentration, and that the very idea of mathematical truth is a concept with some complexity. Obvious, yes, but still mysterious. Mathematics expands its territory of understanding, but it also transforms that territory; it acquires new information and insight, but it is also intimately involved with examining itself.

Moreover, the attempt to make sense of mathematical objects that seemed at first to disrupt the mathematical order was never straightforward. The attitude toward these unruly objects was, at times, a mixture of fear and fascination. But the various and courageous forays led to the calculus, to Cantor's theory, to insights about infinite series of numbers, and to a novel conception of number itself. These attempts to comprehend Proclus's *alogon*—the unutterable—to integrate it fully into a theory of number and space, are part of an effort that continues. Most recently, the late logician Abraham Robinson invented "nonstandard analysis," in which the idea of a literal infinitesimal is taken seriously and is itself made part of an arithmetic, thus deposing Berkeley and restoring notions from the seventeenth century. If the history of this quest for continuity and smoothness tells us anything, it tells us that math is not so much a science as an art.

Abstraction is the crucial process of that art. It is how ideas were increasingly refined, so continuity and the irrational were gradually understood. Each step forward required a reflective step backward, a comparison between what was known and what was shown. New concepts and abstractions were developed, continuing the process and changing the style. When mathematics turns its attention to the properties of a system, seeking to determine relations between its elements and comparing them with others, it must engage in a very particular type of mental activity. That might mean analyzing why one sort of number is different from another, or why numbers act in particular ways in different circumstances. These qualities may then be used to define an abstract structure that may reappear in other contexts, its properties and possibilities replicated in a different realm.

For example, we might create "classes" of positive integers—a division of these numbers into categories—based upon the remainders they produce when divided by 3. These classes of numbers would be

$\{0, 3, 6, 9, 12 \ldots\}$—all positive numbers leaving a remainder of 0

$\{1, 4, 7, 10, 13 \ldots\}$—all positive numbers leaving a remainder of 1

$\{2, 5, 8, 11, 14 \ldots\}$—all positive numbers leaving a remainder of 2

It turns out these classes completely account for all integers. And we can, in shorthand, label them ⓪, ①, and ②, in which the circled number represents a *class* of numbers. We can then define an arithmetic for these classes, under which ① + ② = ⓪ (a number with a remainder of 1, added to a number with a remainder of 2, yields a number with a remainder of 0). Under the same convention, ② + ② = ①. Various properties of this arithmetic can be explored—and we can then define the sort of structure that emerges. That structure can be identified with a general structure of a set with three elements and two operations (+ and ×, usually referring to addition and multiplication) that have certain properties; the same structure can then be found in other cases, some having nothing to do with numbers at all. For example, the musical notes in our twelve-note scale form a set that has the same properties as does the set of classes of numbers based upon the remainders produced when dividing by 12. There are equivalences between the same notes in different octaves: every C on the piano is identified, every G♯, and so on, leaving twelve classes of notes sharing a structure with a mathematical collection of classes. In a broader, more abstract fashion, mathematicians dealing with the issues of continuity were involved in setting up different classes of numbers, types like the "irrational," the "integers," the "infinitesimal." Mathematicians first found differences and similarities; then, over the course of centuries, attempted to refine them.

When we start to think about the processes of abstraction at work in real space rather than in the controlled world of numbers, things get even more complex. Let us return to the torus (the doughnut) and sphere. How are their structures, their similarities and differences, to be explored? We understand something more now about notions of smoothness and continuity, but these concepts don't immediately help us: the torus and sphere both have the same number of infinite points smoothly connected—so how can we distinguish between them? Ironically, the mathematical field of topology, which deals with such questions, emerged only when the priority given to sensory experience began to wane in the nineteenth century and the notions of continuity and number had been clarified. It grew out of the most concrete experience we have of objects and shapes, but quickly left it behind for abstract refinement. It is a field that requires a tremendous amount of mathematical awareness and calls into play many aspects of mathematical practice; it is almost a condensation of different techniques and approaches. It resembles a mature artistic style, in which methods and notions used in previous eras have been absorbed into an overall vocabulary.

Topology asks us to pay acute attention to the intuitions we have developed since infancy. Jean Piaget emphasized that children do not always see the world the way adults do. For example, they may believe that the amount of liquid in a tall, narrow glass is greater than the same amount in a small, wide one, even when they watch the liquid being poured from one to another. They may think a piece of clay is essentially different when it is molded into another shape. Piaget argued that our concepts of quantity and transformation and conservation develop at particular ages. And once they do, we take them largely for granted.

The topologist begins the exploration of space by deliberately refusing to pay heed to the premises by which we ordinarily experience and organize the world. Every question is open, every notion of space and sense rigorously justified. Like the musician who attends to the implications of a composition's chords and their progressions, chords which most listeners treat as inconsequential accompaniment, the mathematician wants everything to be noticed, everything accounted for. This attention to detail, though, is then transcended, for in the detail is found pattern and in the patterns are found similarities. Thus, we may return to the world with a more profound understanding of the sensory experience that we began with.

The notion of continuity we have been tracing, for example, was so hard to formulate mathematically because intuitively it seemed so transparent and obvious to the senses. A sophisticated understanding of smoothness was developed during centuries of analytical reflection, in which different metaphors and images of motion and approach and the infinite were applied and reinterpreted. We can now see smoothness as a kind of equivalence between points, there being nothing to distinguish the neighborhood of one from the neighborhood of another, and no way to separate one from another without disrupting their neighborhoods.

It turns out that the equivalence between surfaces also requires a notion of smoothness. We think of surfaces as equivalent if they can be "smoothly" transformed, one into the other, without breaking or rending or piercing. Such equivalence is based on a sort of rubber sheet idea: if we bend and stretch such a sheet, say, into a hat, it hasn't really changed its essential nature or structure. If, however, we puncture it in the center, it will have been fundamentally altered; it would not be able to recover its original shape or structure. The smooth connection between points and their neighborhoods will have been disrupted.

The actual transformation of one surface into another is

another example of the mapping we discussed earlier—points on one surface are identified with points on another, just as in ordinary cartography. On most maps, for example, the points on a sphere—the earth—are "mapped" onto a finite plane; certain mappings, such as the familiar Mercator projection, must resign themselves to being incomplete, since to include both North and South poles would require an infinite sheet of paper. More simply, a sphere may be mapped onto a larger sphere surrounding it by drawing radii out from the center of each. For every point on the smaller sphere there is one and only one point on the larger, and vice versa. This is a "smooth" mapping we are fairly well acquainted with—the inflation of a balloon. And it helps us understand the notion of neighborhoods of points that is so important, intuitively, to our sense of surface. If two points are close together on an inflated balloon, they can be said to reside in the same neighborhood; the same two points can also be found within a certain distance of each other on a deflated balloon, no matter how small the balloon becomes. A partly inflated balloon and a fully inflated one differ in scale, but inflation does not violate the notion of neighborhoods of any point on the balloon. The bursting of a balloon, of course, would disrupt the neighborhoods completely: a point on the boundary on a burst piece would not have a neighborhood containing a point that was once "nearby."

Something like this is put in mathematical notation as follows: A function inf (for inflation) may be considered continuous at a particular point a (on an uninflated balloon) if for any number ε, no matter how small, a number δ can be found such that $|inf(x) - inf(a)| < \varepsilon$ for all x such that $|x - a| < \delta$.

This language should be slightly familiar from our exploration of the notion of limit and approach, and in fact underlying both ideas are similar requirements about smoothness. In this case it means that if the distance between any two points on the inflated balloon is very small, the distance between the two points on the uninflated balloon is also limited. So a neighborhood on the uninflated balloon will remain a neighborhood on the inflated balloon. No one expects a balloon on inflating or deflating suddenly to change its "character" and become a two-holed doughnut instead of a sphere.

These seemingly esoteric notions give us a method for deciding not just whether a surface is continuous but whether a transformation of surfaces is. The notion of transformation helps us understand the similarity of certain spaces. When there are continuous mappings in both directions—such as with the inflation and deflation of a balloon—the two spaces are called topologically equivalent or homeomorphic. A circle is not topologically equivalent to a line segment, nor is the earth topologically equivalent to any of the usual flat maps representing it, because the mapping is not continuous in both directions. A doughnut is, though, topologically equivalent to a coffee mug.

But even topological equivalence is an inadequate concept for describing the more subtle differences between spaces. It is easy to conceive of surfaces that are topologically equivalent but still differ in fundamental ways. A knotted string tied in a circle is topologically equivalent to an untangled string tied in a circle, for example: the twists and interweavings do nothing to change the "neighborhoods" on the string, but they do, somehow, change its character in ways that need better definition.

The mathematician Henri Poincaré, whose turn-of-the-century inquiries into space have become the foundation of modern topology, defined the problem succinctly:

It has been said that geometry is the art of applying good reasoning to bad diagrams. This is not a joke but a truth worthy of serious thought. What do we mean by a poorly drawn figure? It is one where proportions are changed slightly or even markedly, where straight lines become zigzag, circles acquire incredible humps. But none of this matters. An inept artist, however, must not represent a closed curve as if it were open, three concurrent lines as if they intersected in pairs, nor must he draw an unbroken surface when the original contains holes.

What Poincaré was getting at was more than just smooth transformation or topological equivalence; he was drawing attention to essential characteristics in a space that even the most awkward drawings must take into account: namely the differences that knots and crossed lines and closed curves and holes make. The difference between knotted and untangled circles of strings, for example, has something to do with loops—how we move our pens from one point to another and return to our starting point, or make lassos out of lines. As with spaces, mathematicians explore when loops are similar and when they are different, when they are related and when they are not. The same procedure is repeated: attend to detail and resemblance, find an abstract principle that characterizes those similarities, and repeat the process, always comparing, abstracting, defining.

The loop idea is important in topology because it offers a stronger way of finding similarities than just the idea of smooth mappings; it further refines our notions of space. It turns out that the types of loops and paths that can be made in a space characterize it in a very special way; they present, in a sense, its fundamental phrases, the different ways we move from one place to another.

The concept is intuitively simple. There seem to be possible on the torus, for example, different kinds of loops. We can imagine a loop of string wrapped around the outside surface of the torus (loop 1), and we can imagine a loop wrapped around its "tube" (loop 2); we can also picture a more simple loop on the surface of the torus (loop 3):

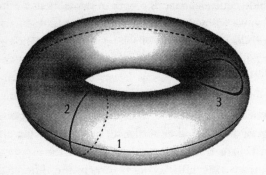

We can see the differences between these loops by imagining that they are created by a small bit of string being held at the starting point by our fingers. Loop 3, on the surface, is similar to the type of loop we can imagine on a sphere: if we were to pull it tight, making the loop smaller and smaller, eventually we would be able to pull the loop into a point and eliminate it. But the two other loops cannot be pulled to a point in the same way. Tighten them and eventually they will be wrapped around either the hole in the torus or its tube. Nor can either one be changed into the other. They seem to have a character that is fundamental for this space.

Once we recognize this kind of difference, we look for its significance. In order to "classify" such loops, there must be some way of defining how two loops might be equivalent: we would want to assert, for example, that a loop around any one tubular section of a torus is equivalent to a loop around another. They are equivalent because one can be smoothly deformed into the other—the loops can be slid over to each other, or, in mathematical language, there is a continuous transformation from one loop to another. Such loops are called homotopic. And in spaces having a certain kind of connectedness without gaps, loops that are homotopic at one point are homotopic at another—so the particular point chosen to make our loop becomes unimportant. We form classes of loops based on this ability to slide equivalent loops together smoothly and continuously, just as we formed classes of numbers based upon remainders they produced and classes of spaces which are topologically equivalent (and, to anticipate, just as in listening to a musical composition we establish classes based upon certain kinds of musical resemblance and equivalence). Each class is composed of equivalent loops.

So instead of being concerned with an uncountable number of loops on a surface, we have, through abstraction and comparison, simplified the issue. We can now attend to a much smaller number of types or classes of loops. On a sphere there seems to be only one such type of loop; it can be pulled to a point; on the torus, there are two that cannot be pulled to a point. So when examining spaces the questions are how many homotopic loops are there and does this quantity give us some insight into the space's structure? The focus of our inquiry has shifted from questions about deformations of the space as a whole— the inflated balloon—to the kinds of lines we can construct on its surface. We have seemingly moved from one level of form to another, as if focusing on the phrasing of a space instead of its overall shape. Or, to put it another way, we are now interested in the way one moves around within a space rather than the way it looks from afar—though it is one of the important insights of topology that these are intimately related.

This movement into a space, though, like many of mathematics's movements into minute detail, has the larger purpose of creating grander views of the space as a whole. We submerge ourselves in detail only to see the whole better, the way we might dissect a musical composition and its premises the better to be affected by its argument. Let us take this just a few steps further, to see how intense this process of submersion and abstraction can become. When we have sensed the importance of loops, we must define them mathematically. Any "loop" can be thought of as a mapping of a line segment—say from 0 to

1—onto a surface, as if the line segment were stretched and wrapped like a broken rubber band so that the end points of the segment—0 and 1—are both directed to the same point on the surface, repairing the band on the surface. A homotopy—which would identify equivalent loops—is actually, then, a mapping between mappings, a function that transforms one function into another. And distinct homotopy classes are actually classes of mappings that are considered equivalent under this extremely abstract criterion.

These are extraordinarily subtle notions, and though they are introduced to every undergraduate math major, they only become clear after much play with surfaces and maps. The levels of abstraction they require are dizzying. In fact, at the most abstract level, we leave behind our ordinary notions of surface altogether; our familiar worlds of space and surface become just special cases for more general structures, metaphors for other worlds. Any set of objects, in fact, may actually be given what is called a topology, a way of measuring distance and relationship between those objects. Such spaces may be beyond our ability to draw or even to visualize.

Still, we can go further and define a way of "multiplying" loop classes, and define a sort of loop arithmetic. When we look at loops using an abstract arithmetic to combine and relate them, we find a startling structure that can have nothing to do with space at all. It is an algebraic structure, a collection of objects and relations that is technically known as a group. The remainder classes I described earlier form a group using addition; so too do organizations of musical notes form groups. The group of loops is justifiably called the fundamental group of a space because it does so much to fundamentally distinguish one space from another.

We can get some feeling of these groups from simple examples. In the Euclidean plane, every loop is the same. Imagine that a piece of string is looped into a circle and held in place by a finger; pull on the string and the circle gets smaller and smaller until it is collapsed to a point. Loops on the plane are thus all equivalent to a point, which is actually a constant loop: it starts in one spot and stays there. Loops around a circle are quite different. It has an infinite number of homotopy classes, an infinite number of types of loops. Each type is defined by the number of times a string can be looped around the circle. None of these loops can be pulled to a point; wrap a loop around a circle twice and there is no way to transform it smoothly into a loop that wraps around it three times. So if the group for a plane has just one element, the group for a circle has the same number of elements as integers. A torus is more intricate: its fundamental group structure requires two "coordinates"—the number of times a loop goes around the tube and the number of times it goes around the doughnut as a whole (see loops 1 and 2). It is equivalent to a set that is the product of the set of integers with itself; its elements are *pairs* of numbers, written like (2, 139).

This is probably far too complex to be easily understood here, but what should be clear is that fundamental groups transform questions about spaces and continuous maps into algebraic problems. Thus, we have a sort of metamapping from spaces into algebras.

Moreover, using this metamapping, we can also begin to get a better understanding of the larger questions. Using fundamental groups, it becomes easier to see how spaces may be transformed into one another, and how they may be related. We step back again from the internal life of the surface into the shape of the space as a whole, and try to find how such a space is related to others. There is, for example, a whole area of study involving the notion of "covering spaces" that asks what sorts of space can be transformed into the one we are studying, in a very special, smooth fashion. A cover for a space can be thought of as a blanket that can model what it is thrown on. It need not even be an unfolded blanket: it can be layered, like a quilt folded over in quarters. It cannot, however, be folded in quarters in one place and in eighths in another, nor can a covering space simply be tossed over a space like the cover on a poorly made bed.

The types of covers for spaces can be surprising: for example, the infinite line is a covering space for the circle; it is simply wrapped around it again and again and again, ad infinitum. The torus may actually be covered by the plane, or by an infinite cylinder (imagine the cylinder twisted and wrapped around in space like the infinite line around a circle). It turns out that there is an intimate relation between covering spaces, their fundamental groups, and the original spaces; the ways in which covering spaces can be constructed are connected to the algebraic structures—the fundamental groups—we have defined.

One last twist before letting this matter rest. The algebraic structure we are talking about is called a fundamental group, but however fundamental it *still* does not completely define differences in spaces. Consider the familiar cylinder and the Möbius strip, which is a circular strip having one side and one edge, easily constructed by giving a strip of paper a half twist before joining its ends. They have the same fundamental group but are different in other respects. This shows how subtle and intricate our naive intuition of space is, and how sophisticated mathematics has to be in order to match our experience of the world. The irony is that once mathematics is powerful enough fully to describe the world we know, it is abstract enough to describe a multitude of other worlds as well.

Topology relentlessly pursues ever more sophisticated and dizzying distinctions of this sort, using concepts of homotopy type, homology groups, simplicial maps, and other exotica. And this is just a glimpse at the elementary aspects of an extraordinarily abstract theory. But the basic approach throughout is plain: if we describe the structure of an object in a sufficiently clear and abstract fashion, we are likely to discover links with other structures. We can then identify the skeletons which lie under their vastly different superstructures. Individual differences become much easier to comprehend once fundamental differences and similarities are known. Differences and similarities are established through mappings, which can even link objects that at first appear to be drastically different. These mappings can themselves become the subject of intense scrutiny. "Mathematicians do not deal in objects," Poincaré observed, "but in relations between objects; thus, they are free to replace some objects by others so long as the relations remain unchanged."

OUR OWN INTELLECTUAL JOURNEY between mathematics and music will be, metaphorically, a mapping of the sort Poincaré described. We are seeking similarities within differences, different levels of equivalencies and relations. We will examine the relations between musical objects and musical thinking, between the minutiae of a composition and the nature of its effect, between musical styles and musical function. The language of mathematics will provide useful metaphors for understanding music. We may find helpful metaphors in our study of number and space—a suggestive connection between our experience of musical space and physical space, how transformation and structure work, how musical "surfaces" are constructed, how we hear music "topologically."

But we are also interested in more than metaphors. A composition can be an exploration of musical material in the same way a mathematical theory can artfully explore a particular space. We hear and understand music in the same way the mathematician understands an abstract terrain. Mappings are made within music—from one phrase to another, from one section to another. They are also made from the sound of the music to our varied experience as listeners. For just as every abstract topological theorem has its mundane special case, which inspired the exploration in the first place, so every abstract musical structure has its particular mapping into the world of each listener. The path we take will be far more abstract than most of us are accustomed to, but only when issues in musical thinking are set forth will the formal links with mathematics become more clear. Then, like the poet, we may hope for some illumination of the surrounding terrain.

III

SONATA:
THE INNER LIFE OF MUSIC

The sense of form of the sculptor, the painter, the composer, is essentially mathematical in its nature.

OSWALD SPENGLER

A T FIRST, AN EXPLORATION OF MUSIC MIGHT BE GREETED AS refreshingly concrete after so much mathematical abstraction. Finally, here is something easily grasped: no fancy theories about continuity and mappings and the number system, just sounds—sounds that are granted to a listener, in W. H. Auden's words, as an "absolute gift," bearing no requirements or preconditions. Who has not felt some inkling of that gift, no matter how crudely offered? Who has not been moved or excited or haunted by its presence?

But this is a most peculiar moment in world musical culture. A great many musical traditions are ending, under siege by the forces of modernity, popular culture, and the entertainment industry. It is a difficult time to contemplate music's meanings. For example, what place does new art music play in our lives? Could a contemporary opera composer be, like Verdi, hailed by ecstatic crowds outside the theater after a successful premiere? Only pop music generates such enthusiasm, but pop music fulfills a different function from art music and often has different ambitions. As for folk music from around the globe, it has long since begun to loosen its connection to premodern life and is increasingly subject to dislocation, transformation, and, in some cases, extinction. Béla Bartók traveled around Hungary at the beginning of the twentieth century attempting to record folk songs that, even then, were being forgotten. Today, on the eve of the twenty-first century, the transition to a global musical culture is nearly complete. Much folk music is now called world music.

Today, music is ubiquitous, utterly entangled in our daily lives. Never before have so many different kinds of music been heard by so many people so often. Music is unavoidable, the boundaries between sound and silence, between occasion and accident, now thoroughly permeable. Music bombards our ears in advertising jingles, popular hits, round-the-clock music stations. It spills out of concealed speakers in elevators and on trains, in airports and in shopping malls. We do not need to search for music; it seeks us and finds us out.

So we cannot complain, as we might of mathematics, that our sense of music is thwarted by inadequate exposure. All kinds of music flow together today in an uncharted stream, currents hardly demarcated by goals or methods, swirling into a vast, undifferentiated pool. Notions of beauty and truth and knowledge dissolve into assertions of taste, social function, preference, and habit. There is, of course, a venerable tradition in which music is thought of as little more than pleasurable sensation; at the end of the twentieth century, that is generally all most people require of it.

This pleasure is often rooted in the listener's sense of "identification" with the music: feeling that Bruce Springsteen, say, or Hector Berlioz understands and precisely expresses the listener's inner, unarticulated (but deeply felt) sentiments. For many people, music is most successful when it mirrors the listener's state of mind, or when it manages convincingly to create one. The composer, in this view, tries to express in music emotions such as sadness and anger and happiness and fear; we in turn imagine we feel these sentiments when the music is played. This was the vision that came to maturity in the Romantic era—put forth by C.P.E. Bach and turned into the vocation of "self-expression" by Liszt, Berlioz, and others, who paraded their dreams and passions across the concert stage.

But does music really represent feelings in this way, either by reflecting our own or by expressing the composer's? Is it emotion we feel when we listen—anger or sadness or envy or desire? When music brings tears to our eyes, is it because it makes us sad? Some music unquestionably does stir or inspire us; that is the purpose, after all, of national anthems and masses and even some folk songs. Some music also prompts unexpected emotion and thought. But this view of music's purpose is far too limited. The Indian raga serves the function of neither pleasure nor expression, nor does most of the great music of our Western tradition—even the Romantic music that claims to be fundamentally self-expressive. Only products of the pop culture industry unambiguously aim to inspire identification with musical "expression," and seek the avid consumption of such expression through purchase or use.

Our task is to probe more deeply, to try to understand music's methods and power while acknowledging that notions of identification and self-expression are inadequate. One obstacle is the overwhelming variety of music. Is there anything we can say about Ars Nova of the fourteenth century that would also be true for, say, Stockhausen's twentieth-century experiments with extravagantly staged musical theater? What, aside from the brute facts of physical vibration, does Balinese gamelan music share with Bach's cantatas? Or Wagner's operas with synagogue chant? In what way, if any, is the history of musical invention similar to the history of mathematics, in which styles change but truth seemingly does not? Music varies so widely from civilization to civilization, from culture to culture, from time to time, even from moment to moment, that it is tempting to embrace a bewildered disinterest about its intentions and methods. How far back should we stand when looking at the phenomenon of music? It is probably foolhardy to look for fundamentals given the welter of variation; we may have better luck by examining the differences between kinds of music.

Such differences, after all, direct our attention to essentials, for they are far from arbitrary. Music takes its shape and style for any number of reasons—the composer's personality and intellect, the function a composition is meant to serve, the audience or occasion for which it is written, the attitude of the composer to that purpose and audience, and, not least, the nature of the musical tradition out of which the music is written. It may be that one of the essential characteristics of music is that it allows such connections—such mappings, if you will—to be made.

For example, there are aspects of music that depend solely on the purpose to which it is being put. Some of these may be obvious. If music is meant for dancing, the fundamental interest is in rhythm. In cultures where dance and ritual and trance are important, the repetitive rhythmic organization of sound is essential for closing off all other senses of period and temporal division. By contrast, in cultures, such as ours, in which "social" dancing is common, in which the action of the dance is meant to be a gracious sort of conversation between bodies through space, the rhythm must be firmly articulated (or playfully unarticulated) so that the motion of feet and bodies can be felt in the swing of accent and release. It must only be startling in ways that will support the powers of dance—its surprise should inspire physical exuberance rather than shock (except, of course, in the most contemporary of dance styles). But any intricate counterpoint, any indication that there are many voices in the music, each acting independently of the others as in a Bach fugue, would simply confuse the attention of the dancers. One of the great achievements of Bach's dance music, for example, was to be contrapuntally crystalline while maintaining the subtle press and pull of the dance.

A function like dance acts as a sort of lens for a style: it magnifies certain aspects, and helps to define others. This can even be true of more subtle functions. Consider an obvious example: music written for outdoor performance must be played on instruments that project across large distances; violins are too thin-bodied, flutes too airy—brass is ideal. Moreover, the music's organization must rely less on subtle inflection than on broad gesture; it willingly sacrifices intimacy. It is no accident that Handel's *Water Music* sounds constrained and obvious in a concert hall but would sound exhilaratingly playful if heard by a river, as it was when first performed. Concerts played in parks with artificial amplification should not attempt to include the classical symphonies that were written to be heard in the acoustically sophisticated halls of Europe—the supporting lines are relegated to background hum; the music is reduced to familiar themes and blatant rhythms. Brass bands would have greater success. It even makes sense to talk more generally about "indoor" music and "outdoor" music. For example, the Chinese pipa is a lute that was once played in the traditional courts; the gamelan music of Bali with its robust interweaving of rhythms is traditionally part of outdoor rituals.

We can make another sort of large-scale distinction: we have little difficulty sensing when music is written for a public occasion—the celebration of a mass, the commemoration of a battle, the honor of a citizen—and when it is written for private consumption. Benjamin Britten's *War Requiem* and Bach's *Well-Tempered Clavier* are not just vastly different in their musical language; they are meant to accomplish different things, the first something public, the second something private—perhaps something solely for the player. Public music is, to a certain extent, simpler than private music. It must be clearly grasped by a wide variety of listeners at the same time. It must use broad strokes—like outdoor music—and it must assume a shared interest and occasion. Private music can be more idiosyncratic; it can presume less and offer more. The effect of recordings in the twentieth century has been to break down these boundaries radically—since all music is now potentially private—with drastic effects on composition and sensibility.

We can make more subtle distinctions, beyond simple functions, about the differing natures of religious faith and belief, and their effect on music. In traditional synagogue chant, for example, can be heard a representative cry of people whose voice is singular rather than choral, and whose path through history is represented by the melancholic twists of melodic lines. By contrast, in many settings of the Catholic Mass may be heard an attempt to transcend the forces of history altogether; the suffering of the individual voice is absorbed and transformed into the choral proclamations of the faithful.

I am not outlining something esoteric here in describing the reasons for musical variety but something fundamental. It is the way music may be said to "work"—the way it is applied to a situation and serves a function: the composer finds out what is needed and why, takes the material given in the art, and attempts to find an appropriate "model." This modeling involves finding a musical metaphor for the social or religious or expressive occasion, tones mirroring the relations found outside them. This is a subtle issue, but it is one we may begin to understand in an obvious case—music written for a film.

During the era of the silent film, a pianist or an organist was often stationed in the theater, moving eyes back and forth between the screen and the keyboard. Before him might be a rather fat book, such as *Motion Picture Moods* by Erno Rapée, in which "genre" pieces were indexed according to dramatic function. There were pieces appropriate for chase scenes, for moments of suspense, for lovers' reunions, for villainy. Each of these pieces—some taken directly from the classical repertoire and thus familiar to the audience—captured in some way the gesture being displayed on the screen, the way something of the headlong rush of the Lone Ranger is captured in the overture to Rossini's *William Tell*. These are, in other words, not accidental conjunctions of image and sound but examples of musical modeling.

Such modeling has to be accomplished in film or dramatic accompaniments so accurately that the music never draws attention to itself; it serves simply as a form of punctuation, pointing a finger at an aspect of the scene appearing on the screen, manipulating our response. The scoring of films has grown extraordinarily more sophisticated since those early days, but the function is the same. The music is a guide to what is being shown. The cracking of an egg can be a dramatic turning point, a mundane act, or a gesture creating excruciating suspense—depending on the music accompanying the image. There is even something Baroque about the notion of film scores; they are like early operas, in which the music is used to support the stylized movements and gestures of royal figures and lovers. During the seventeenth century, for instance, it was felt that every composition should aspire to no more than a single affect: it should be weeping or joyous, fearful or exuberant, but not involve multiple sensations. The transitions between these affects were never taken as seriously as they would be by the Romantics more than a century later. Film music is a form of just such musical punctuation: the less ambiguous it is at a certain moment and the more single-minded, the better (unless, that is, ambiguity is the main dramatic point). This doesn't mean that it must be crude or brutish: it is even possible in this context to present music that is a lot more daring and difficult than might otherwise be attended to in a concert hall. An audience who would squirm in their seats at the dissonances and agonies of the Chamber Concerto of Alban Berg would be on the edge of their seats were the very same music to accompany an Expressionist horror film. In fact, Arnold Schoenberg, who almost cultivated audience squirming with his music, tried to write movie music when he found himself in exile in Hollywood in the 1940s.

The modeling I am describing is, of course, not a matter just of "style" but of the way style is used. We can have "suspense" music in a Baroque style (think of the harmonic patter of a Bach harpsichord concerto, the teasing, climbing repetitions of short fragments) or in late Romantic style (what is the music of *Tristan* but a four-hour study in suspense and anticipation?) or in an Indian raga (when will the rhythmic and melodic meditation return to its origins, after what gestures of accent and variation?). We can also have outdoor music or religious music or private music in any number of styles; it may even be possible to accomplish something similar in totally different styles. So the very notion of style needs better under-standing. It may be, as in mathematics, something that is always implicit, never explicit, the presumptions and assumptions, most barely evident, of how music is to be made and what it is to explore, what vision it holds of itself, and what of the human mind. A musical style might be, as in mathematics, a way of getting at the truth in a particular time and place.

It is difficult to try to grasp something so elusive, and, by definition, barely visible. But it is something we are always attending to in that half-conscious way in which we often listen to music. Just as the mathematics of Leibniz's day and that of Cantor's differ in kind and style as well as knowledge, and just as these differences are immediately apparent to anyone who begins to try to think through arguments made in different historical periods, the moment we try to listen to music with any kind of attentiveness, differences in style present themselves as self-evident. Musical styles effortlessly communicate to our listening ears all kinds of presuppositions and attitudes, matters which may never be explicitly articulated but which are nonetheless always felt. The highly educated listener can hear a few bars of a composition and make a plausible guess, within a decade, when it was probably written and where, even if the music has never been heard before—the way anybody who grew up with today's popular music can date its styles by year, or the way we can glance at a photograph and know when it was taken by the way hair is combed or clothes worn. The stylistic elements are so ubiquitous, and our perception of them so immediate, we may not even be able to specify what it is exactly that gives us these cues, yet every musician knows them and lives by them.

A musician sits differently when playing Bach at the piano than when playing Brahms. The hands in Bach are like complex organisms, each finger having its own independent life but communing thoroughly with its neighbors; for Brahms, the hands are less communities of fingers and more extensions of the body: they are made to seem thick and broad, linked to the arms and shoulders and back. Physical pose is also connected to the sensuous spirit of these two styles, the first encouraging an ecstatic distance, the second a physical brooding. Similarly, musicians know that when singing the music of, say, Josquin des Prez, the expressive gestures of the voice are different than they are when singing Verdi: in Josquin's work a crescendo is not a climax of passionate feeling but a sign of something barely visible under the surface, an affirmation of nearly melancholic devotion.

So when you play music, you also embrace a style. A style suggests ways to sit, ways to sing, ways to feel rhythm. It also suggests ways to think. A style even defines a musical community—a group with shared notions about music and its purpose. The shared style allows for musical communication without misunderstanding, a common sense of what is being said and why. The same style of music presented to different communities will create different reactions, just as Expressionist music would if played in a horror film, a concert hall, or a Catholic church.

It is harder to give examples of musical styles than of mathematical styles. In mathematics a style is obvious when the same fact is proven in different ways; in music style is just as fundamental but more elusive. Still, we can get some sense of style from melody alone.

Here, for example, are four melodic lines. The first is a notated version of the traditional chant used to recite the Hebrew scriptures in synagogues. This chant influenced the development of early music in the Christian Church. Here is how the first phrase of Genesis might be chanted in Hebrew:

BE-RE-SH-E ET   BA-RA   E-LO-HIM   ET-HA-SHE MA-A-YIM VE-ET   HA-A-RETZ

One traditional melody for the chanting of the first sentence of Genesis in Hebrew. Upper level commas mark word breaks.

The system of chant is set up so that each word is linked with a set of notes, a trope. The melody is indicated by a sign which appears above or below the Hebrew word. The tropes are themselves linked into phrases: they are often grouped in repeated patterns associated with the phrasing of the Hebrew sentences. They are literally a heightened form of speech, serving as a punctuation for the text. Musically, the punctuation of greatest importance is the falling line that marks a sentence's end.

The "style" of this chant is built out of a limited number of musical elements, the few dozen signs or tropes generally presented in standard groups. A particular set of tropes can generally follow any other phrase; in fact, except for a few circumstances, the tropes combine into units which point neither backward nor forward musically. Each time a sentence ends a new one begins, unaffected by the music that preceded it. The style is not developmental but static. Its expressive quality is due, in part, to the linking of unchanged musical patterns with timeless words. There are no ambiguities in the line, no complications in which a musical tension would distract from or comment upon its text. The music is generally not meant to illuminate, merely to punctuate and emphasize.

Now look at the musical style of this violin line from Bach's *St. Matthew Passion*:

Violin solo beginning the aria "Erbarme dich" from J. S. Bach's St. Matthew Passion

Here the relationship with the text is entirely different. The melodic line's function is not just to punctuate but to interpret, to illustrate. It is a musical image of its subject, a plea for Divine compassion. The particular tones imply a network of connections with their predecessors and successors. This phrase cannot stand on its own. It is left unresolved—open—demanding fulfillment. In this it is also expressive of an entirely different religious universe as well as an entirely different musical one. It is part of a drama in which a fragment is insufficient for full meaning. This presumes a sacredness not of the text, as in the reading of the Hebrew Torah, but of the ideas the text expresses; the words are not sufficient, they are only tools.

These two examples are extremes, of course, but their styles determine many things: what sorts of notes are possible, what settings are permitted, what the function of music is to be, even how we are to listen. We also don't require a text to feel this essential connection between style and meaning. Here, for example, is the theme upon which Mozart began a set of variations in a piano sonata:

Opening phrases of Mozart's Sonata in A (K. 331)

If we wanted to cut this theme off at some point, we would be left with a feeling of incompleteness. Each musical gesture is answered by another: the first two measures by the second two; the first four by the second four; the first eight by the second eight. A notion of symmetry and balance comes into play here that involves rise and fall, tension and release. Even the rhythmic motion has a regularity and a pulse that changes, with calculated elegance. This is a form of rhetoric. If the number of measures were changed, if the melodic line suddenly took an unexpected leap, if some bizarre musical event were left hanging without response, the style itself would be violated—as it was, with deliberate expressive intent, by later composers.

Consider finally a melodic line from Anton Webern's Second Cantata:

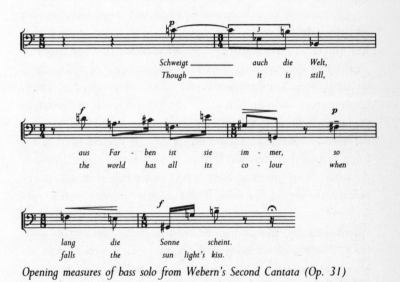

Opening measures of bass solo from Webern's Second Cantata (Op. 31)

At first glance, this line has completely different presuppositions about what musical gestures are permitted and how they should be combined. The musical universe is radically different: we cannot find in this statement an obvious regularity or balance. That is, of course, one of its purposes—to encompass great reaches of musical space, leaving implications as free floating as objects in a world without gravity. The conception is different along with the intention.

These sorts of differences are as profound as they are obvious. A musical style not only defines what is permitted in music—

whether a particular chord would be appropriate in a particular place, for example—but also creates a hierarchy of musical value, determining what is important and what is irrelevant—whether melody takes priority over rhythm, or whether one type of rhythm supersedes another. Like a mathematical style it gives a musical project shape and direction. It might require that music serve a particular function, but it also ensures that music will serve that function in a particular way.

One reason Western music has had such a powerful impact on the world and why it is so sophisticated an achievement is that in it the function of music—whether used for ritual or dance or worship or pleasure—gradually became an *aspect* of style, not its defining force. Dance music could stand independent of dance; church music independent of the church; art music independent of patron—because the internal world of each composition was sufficiently coherent to stand on its own. It had no need to remain tethered to the world that gave it birth. Once music is detached from function, once it becomes a repertory art, it explicitly strives to define itself, out of itself, to become "mathematical"—that is to say, to begin from premises and proceed to conclusions by interpreting its own universe, finding its own laws. Systems of harmony and counterpoint become tools for elaborate musical explorations. A great deal of Western music is as much a manifestation of idealism as is mathematics.

THE CRUCIAL MOMENT in the history of idealism in Western music was the emergence of the Classical style as embodied in the compositions of Haydn, Mozart, and Beethoven. The Classical style is also the most refined example of a style as I have been describing it, for one of the goals of this music was to create the sort of internal coherence we have come to associate with the very notion of style—an argument made brilliantly in Charles Rosen's *The Classical Style*.

Crucial to Classical style was the sonata form. Every student of music learns a definition of sonata form that treats it as a formula for composition—a kind of pie plate or baking mold shape into which musical material is poured. The sonata, in this definition, requires three sections—an exposition, a development, and a recapitulation. It includes, according to Donald Jay Grout's textbook definition:

> (1) an exposition (usually repeated), incorporating a first theme or group of themes in the tonic, a second more lyrical theme in the dominant or relative major, and a closing theme also in the dominant or relative major—the different themes being connected by appropriate transitions or bridge passages; (2) a development section, in which motives or themes from the exposition are presented in new aspects or combinations, and in the course of which modulations may be made to relatively remote keys; (3) a recapitulation, where the material of the exposition is restated in the original order but with all the themes now in the tonic; following the recapitulation there may be a coda.

This definition makes the sonata form into a mannerism—a mold rather than a style—following rules for composition the way a waltz follows rules for rhythm. But, as Rosen argues, "The 'sonata' is not a definite form like a minuet, a da capo aria, or a French overture: it is, like the fugue, a way of writing, a feeling for proportion, direction, and texture rather than a pattern." Rosen suggests shifting the emphasis in our understanding from a set of rules to a notion of tension and balance. If, during the Baroque period, the energy of a work, its motoric principle, came largely from the sequence—the repetition of patterns at different pitches creating or lessening tension—in the Classical style the energy comes from the articulation of dramatic conflict. Listening to a Classical sonata, one hears statement and counterstatement, tension and reconciliation. The Classical sonata is ideally a form of argument, growing out of the material with which it begins. "A classical composer did not always need themes of any particular harmonic or melodic energy for a dramatic work," observes Rosen. "The drama is in the structure." Rosen goes on to quote Friedrich Schlegel, who expressed the changing vision of music at the end of the eighteenth century:

> It generally strikes many people as strange and ridiculous if musicians talk about the thoughts in their compositions; and often it may even happen that we perceive that they have more thought in their music than about it. Who has a feeling, however, for the wonderful affinity of all the arts and sciences will at least not consider the matter from the flat and so-called "natural" point of view, according to which music should be nothing more than the language of sentiment, and he will find a certain tendency of all pure instrumental music to philosophy not inherently impossible. Must not pure instrumental music itself create its own text? And is not the theme in it developed, confirmed, varied, and contrasted in the same way as the object of meditation in a philosophical series of ideas?

Our task is to try to understand how such a theme is developed, confirmed, varied, and contrasted as if it were an idea without getting involved in knotty issues of theory and labeling. What sort of thinking goes on in such a style? Is there any way to move beyond the notion of style to the way music works, the sorts of knowledge it gives us, and where it can lead?

In mathematics we have some idea about how things work. With more exposition we could arrive at a clear sense of why mathematics developed, what its procedures are, where it can lead. Different styles of mathematics add to knowledge and understanding. Mathematics can speak about things: it posits them and sets up relations between them. But leaving aside "pictorial" or "program" music—the sounds of birds singing or the sonic imitation of moonlight glinting on the waves—

music does not seem to create or resemble objects we know at all. It never says, as would be said in mathematics, "Let $\mathscr{H}$ represent a two-dimensional Euclidean space." What music "says," it simply shows. It doesn't seem to be able to talk about itself, or to talk about what it is doing.

In mathematics, the process of modeling—of distilling essences, recognizing similarities, articulating principles, drawing conclusions—is essential. We may begin with our experience of the world; then we abstract from it, defining structures and relations. We map from structure to structure, creating new ones along the way. The very movement of mathematics, its progress, can be connected to the tension between such models and the world they are meant to model. I used as one example the fundamental distinction between our experience of the world and our experience of numbers—the contrast between continuity and discreteness—and the various attempts to reconcile the two in notions of numerical and spatial smoothness.

But in music neither the model nor the object nor the map is clear. How could music possibly progress in its understanding of a concept or an experience? I spoke metaphorically about film scores "modeling" an emotion and about a piece of music serving as a "model," but what can this mean? Music does not even seem to be looking for something to model from the world; nor does it seem to involve the sort of reasoning we find in mathematics. In music we don't see the act of construction taking place. We don't hear the premises being set or the conditions made, or see arrayed before us justifications for each step. We are not proceeding from the unknown to the known or from the known to the unknown but are submerged in a realm in which, at least at first, "knowing" seems to be irrelevant.

But to a certain extent such ideas mistake the activity of mathematics with the product of that activity. As we have seen when we look only at that product—the neat mathematical conclusions and statements about spaces and surfaces, for example—we do not fully see the process. We see something that looks like a well-built sculpture, each part fitting clearly and cleanly into the next. Sometimes the product—the result—starts us thinking about how it was reached or where it is reaching, but only when we look behind the product do we begin to sense the inner life of mathematics.

In music, though, the product *is* a process: it occurs through time and in time. Something happens in music and through music. A composition changes us as we listen. It may be that the process of doing mathematics should be compared with the product of musical thinking—the music itself. This makes sense, for there is a way in which music teaches itself. It can be learned simply through listening. Something in the music cues us how to listen, then lets us realize what we are listening to.

Mathematics may appear more inherently difficult than music, but music's methods are far more puzzling. When we paid attention to differences in musical style, we saw, however tentatively, that some kind of modeling does go on in music—some sort of construction that is meant to bear a certain relation to our emotional or physical or intellectual life. There is a relation, often quite crude, between technical aspects of rhythm or form or melody and the music's function.

The way this occurs is not simply through systems of counterpoint and harmony. It is quite possible to be acutely sensitive to music's explorations and discoveries without being conscious of a particular harmonic progression or identifying what species of counterpoint is being used and why. Indeed, these systems have often become obstacles to musical understanding. If the naming of harmonies and types of counterpoint becomes the focus of attention, then we are missing the life to which these labels apply. These systems provide only a lexicographical metaphor for music: they order and list. A question like "What is this piece about?" is answered by saying that it is an exploration of the "Tristan chord" or a study of the Neapolitan second.

Naming keeps music's relations and constructions within a tightly ordered closed system.

But as we have seen, there is a life to the ideas within a mathematical proof, a tension between something known and something yet to be; and often a sort of wit that surprisingly connects what seemed unconnected. Mathematical "movement" is not in individual statements but in the relationship between statements. The concept of mapping is so important in mathematics because it is itself a model of mathematical activity. A mapping is between things: it is the life of argument, linking and identifying and relating. The life of music may have some relationship to this concept as well: how else would music be able to teach itself, unless there were a process at work that made connections, showed similarities, identified differences, and conveyed their significance? The naming of chords must be in service to this other concept, to our sense that music is not about things but about the *relations* between things.

Consider the melodies from Bach and Webern I just discussed; though their styles and premises are distinct, we had to approach them in similar ways; we had to think about what sorts of forms they took, how parts of the melodies seemed to relate to one another, and how these relations might tell us something about the melodies' purposes and presumptions. Consider, in more detail, a simple melody, one that is bound to be familiar: the opening of the "Marseillaise."

Beginning of the "Marseillaise"

This is an example used by Victor Zuckerkandl, a Viennese philosopher of music who wrote three volumes attempting to answer some of the questions I have been posing; they are unassuming volumes, almost elementary, but there is a subtle power in the notions they introduce.

The music setting the first line of the text (from *allons* to *patrie*) in the opening phrase of the French anthem uses just ten tones. Yet the effect of these tones is immense, and it is not solely due to the politically charged associations the anthem has had over the years; the tones are inherently proclamatory, raising expectations. There is a thrust and upward movement in the line, which reaches to that high E from the lower E. The rhythm propels one upwards because it does not begin on an accent but leads toward one. The opening tones are like the pulling back of a fist about to strike. They are also a model of the thrust of the overall phrase; the high held E becomes both exclamation and challenge.

But Zuckerkandl asked a question at once more obvious and more subtle. Where does the sense of motion and thrust come from in this melody—or indeed in music as a whole? We speak of music as moving and often define that motion in extraordinary detail. It is not, though, the motion of tempo and speed—it's not a matter of how fast music is played. Nor is it motion as we usually think of it. A melody is composed of discrete notes. They do not slide into one another. They may overlap, but they also may be completely separated. In any motion there is something that remains the same but changes its place, but in music what remains the same and what moves? As Zuckerkandl wrote: "To hear tones means to hear nothing but tones; besides the tones, there is nothing else, no background, no frame, before which they might move."

Where then could the sense of musical motion come from and what is its importance? There is certainly a change in the frequency of tones heard—the pitches of the melody are varied. But the movement of pitch is itself curious. Our sense of musical movement is continuous while pitches change in a melody by discrete steps. The movement from A to E is a leap. Yet somehow we experience it as a continuous movement. If we attempt literally to connect these tones, to fill the spaces between them, we get no more than a slide or a siren—a sound which is almost musically irrelevant and can also seem oddly static.

Perhaps musical motion comes from meter and rhythm—the accents we feel when we sing the French anthem's opening phrase. In music there is a division of time; each note is placed in a particular temporal spot and has a determined duration. In Western tonal music, there are generally equivalent divisions of time called beats, which serve as the sort of frame Zuckerkandl was looking for. A note can fall off the beat or on it; it can disrupt the beat or support it. There are also groupings of beats into measures, which literally measure out the time by giving similar importance to beats at a given regularity. Every third beat might be accented, for example, creating a waltz rhythm. And every third beat becomes a new start for the pattern: hence, we count 1, 2, 3, 1, 2, 3, 1, 2, 3.

But as with discrete tone, so with discrete beats—a rhythm is not like a sequence of numbers at all; it is closer to our experience of continuous time. When we feel rhythm subtly, it is not like the thumping of a mechanical drum machine, with accents calculated and then routinely repeated; it is more like the movement of a conductor's baton or Fred Astaire's feet. The model for rhythm is not the goose step but the breath—the inhale and exhale—or the heartbeat, with muscular contractions of interior chambers. This sort of rhythm slides and elides. "Musical meter is not born in the beats at all," Zuckerkandl concluded, "but in the empty intervals between the beats, in the places where 'time merely elapses.'" Meter is continuous rather than discrete. Measuring it with a metronome is like considering an arrow to be at rest at every moment of its flight because we can specify its location.

This conundrum is similar to the problem we saw in mathematics between the continuity of space and the discreteness of number. Only here the problem is the continuity of musical sensation and the discreteness of tone and beat. In music, of course, this tension has a different importance, but it still points to something fundamental: how music uses its divisions of musical space and time to create the palpable sense of continuity, how a series of discrete tones becomes a model of something quite different, something Hegel recognized when he wrote of music's "echoing the motions of the inmost self," its power "of penetrating with its motions directly into the inmost seat of all the motions of the soul." Musical knowledge, in some way, is a knowledge about the kind of continuous musical space that can be created with discrete tone and beat—as if we were attempting to model continuity with discontinuity. Music's great energies derive from the creation of continuity out of discontinuity—a sort of inversion of the calculus, interested not in the infinitesimal and the instantaneous but in the ways they combine into the gestural and fluid that resembles in some inchoate way our inner life.

This is as literally true as a figurative truth can be: melody and meter are forms of gesture. Just as a gesture has a nonverbal sense that cannot be easily articulated, just as it has a continuous movement, making it single rather than multiple, just as a gesture uses the body to create a motion with a particular linguistic meaning quite apart from the physical—so too in music: it is nonverbal, continuous, physical, linguistic. Continuity comes from something beyond literal pitch, found in the very notion of relation. Just as the mathematician can see a number in terms of its nexus of relations to others—its properties granted by nature, and its place in a particular organization of concepts and rules established by the art—so does a musician experience tone. Melody is a musical state in which these relations are harnessed, creating a field of tensions and relaxations, anticipations and surprises.

The tonal music of the West is one of the most highly

developed systems for organizing these tensions and relations. It is a kind of musical geometry, articulating relations of closeness and distance, curved lines and intersections. And it is found in this most elementary form of relation—two notes defining a kind of force field in musical space. Nothing marks an "unmusical" presentation of tone or rhythm more than the failure to attend to gesture and continuity. Listen to a child play note by note at the piano; then ask the same child to sing the melody. In most cases, the singing will reveal a sense of continuity which is lacking in the playing. You can define a "musical" performance immediately by the way in which one note affects another. The motion that Zuckerkandl was attempting to examine was not that of an unchanging object from one place to another, or of a changing object seen in different places. It is the change in the network of forces created by tones—a network that extends even to the spaces between notes and is felt and understood as gesture.

Zuckerkandl illustrated this with an element of Western music as simple as the major scale. Habit may prevent us from being too aware of what we sense in this scale, but if we sing it slowly, and attend not just to the notes but to the sense of gesture they suggest, the scale takes on another dimension. It seems first simply to rise—do-re-mi-fa-sol-la-ti-do (①-②-③-④-⑤-⑥-⑦-①).* But ears trained by tonal music will hear something even more subtle in the scale—what Zuckerkandl called its dynamic field.

Hearing that field means feeling the scale as a movement toward a goal—a departure from ① and an "ascent" to it. The scale sung seems complete, with all tensions resolved, in part

---

*Since the scale can begin on any pitch, we will represent its note by circled numbers. Any C in a C-major scale will be ①; and G will be ⑤. This will also emphasize that the important issue is not pitch but the relationships between pitches. And it echoes the approach we took in referring to classes of numbers.

because the final ①, though "higher," is felt to be equivalent to the first. This is the principle of octave equivalence. Recall: if the vibrating string is divided into two parts, or four parts, or—to be mathematically precise about it—$2n$ parts, consonances are created that we often identify as equivalent to the way we identify equivalent hours of different days.

But if the scale is felt as a departure from and a return to that all-important primary note, at what point does it shift direction? Here again we go by impression, by a musical sensibility derived partly from nature and partly from centuries of cultural application: the movement from ① to ⑤ is an ascent felt as a movement away, requiring an expenditure of energy, like rolling a ball uphill. After ⑤ something strange happens. It's not as if the movement completely reverses, but it changes character, as if ⑥ is both an effort and a release; by ⑦ the ball is nearly rolling on its own down the musical hill. By ① we have returned to the same spot (or a corresponding, higher-pitched spot), moving in a spiral if not a circle. The scale in Western tonal music, then, has a peculiar shape and energy. It is far from symmetrical, and its nuances of tension and relation become more intricate the closer we look.

Any tonal melody we sing exploits some of those tensions and implies others which grow out them. Consider again the thrust of the "Marseillaise." It plays with the maximum tension and distance between ① and ⑤—between the first and fifth tones of the scale. It wouldn't have its initial energy if it didn't begin forthrightly at the point of maximum distance—on ⑤ (for "*allons en-*")—but with a weak accent so when it lands on the A—the ① note ("*-fants*")—it is propelled again toward the higher ⑤ from which it later descends. The melody is like a pendulum whose swing begins by being lifted to its maximum height and is then released. But, when we look closely, we see that the melodic gesture ⑤→①, which begins the theme, is followed by ②, the second degree of the scale (setting "*la*

pa-"), which would tend to lean backward to
in the melody instead is that the musical gest
to ① (the interval of a fourth) is repeated from
moving from the B to the E in the first two sylla
tri-e." This repeated gesture seems to increase the melo   st,
as it carries us upward. It is a gesture of triumph.

Baroque fugues make these sorts of melodic echoes and
tensions particularly clear. In contrapuntal music we are attentive
less to large-scale structural events than to the warp and pull
of individual lines, plying their intervallic explorations. They are
made all the more powerful because they often act as if they
were oblivious of one another, as if their only concerns were
internal—the way a melodic leap is made or the way tension
is released or sustained in a few successive turns. For instance,
the Bach fugue is elemental: it may build to a climax by exploiting
harmonic transformations, but what it celebrates is the individual
voice communing with other like voices in a musical society.
The character of that society is determined by the character of
the theme itself; Bach was able to glance at a theme and tell
immediately how it should be treated in a fugue. The fugue's
power comes from the power latent in the melodic field created
by the tones of the theme.

Consider again the theme of Bach's D♯ Minor Fugue (from
Book I of *The Well-Tempered Clavier*), one of the most haunting
of all of his fugues because of both its contrapuntal invention

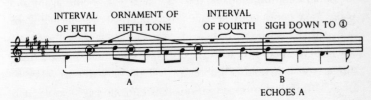

Theme from the first three measures of the D♯ Minor Fugue from Bach's **The
Well-Tempered Clavier**

and the nuances of the theme itself. Those nuances, it turns out, are affecting because they are so simple. The two most important notes of the theme are the first two—the first tone leaping to the fifth. This becomes a recurring motif. It is a form of horn call that does not climax in a proclamation but expires in a sort of prolonged sigh, releasing the energy within the interval of the fifth. First comes a caress of that fifth tone, a circling around it that seems almost to survey the surrounding tonal space from an elevated platform. Then the initial gesture is repeated in a different form: it begins with a leap that leans on the fourth tone of the scale, and releases its tensions, step by step. The entire fugue depends on these two leaps—the first to the fifth tone, the second to the fourth. The first is dizzying and bright; the second, partly because of its shorter goal, is already a sigh of gradual relief and descent to the stable first degree of the scale. These leaps are heard again and again, at times overlapping each other, leading to the most extravagant contrapuntal play: the melodic line is doubled in length, or turned upside down, played against its inversion or, fast on its own heels, in a fugal stretto.

The fugue manages to create a grandly scaled musical space explored with as much energy and invention as the science of Bach's time was beginning to explore earthly space. There are even different surfaces constructed within this musical space. At times the theme enters with lengthened notes, so that one voice intones the theme in extended, nearly suspended time as the others continue unaffected. We hear the theme again and again as a continuous surface, suspended above or below or between the others.

The creation of these musical surfaces depends on something similar to what holds a bubble or a drop of water together— a sort of musical surface tension. It is enormously flexible; it is as dependent upon the skill of the performer as on the talent of the composer. But creating and re-creating it requires understanding the nature of melody itself, its tensions and axes and echoes (the two leaps of the fourth in the "Marseillaise" or the fifth followed by the fourth in the Bach fugue).

A "variation" in music retains the structure of such a melodic line—its tensions and relations—while playing with its inessential elements. In the midst of that play, aspects of the melodic structure are illuminated as if in different lights. Mozart, for example, took the melody we now know as "Twinkle, Twinkle, Little Star" and created an ebullient display of patter and poise in a set of variations, never leaving the original melody behind. The essence of that melodic line is its "argument," around which any number of changes can be rung. To a certain extent, entire compositions can be viewed as elaborations of a melodic line—extensions of simple movements over vast periods of time, with subsidiary movements and tensions appearing under the arcs of the larger movements. Thus, melody can be suspended in musical space. Our ear can retain the sense of that suspension; like a mathematician unraveling the long line of a complicated argument in a proof, we can feel melody's return and resolution.

This sense of melody suspended over time, of tensions left unresolved and picked up later in order that a surface may be created over the length of a composition, is fundamental to our musical hearing. It is what allows us to speak of suspense in music and what constitutes, in part, the integrity of a composition. This notion of suspension, of extended surface, is one way tonal music addresses the passing of time itself.

Bach was expert at this: part of the genius of his counterpoint is that it creates multiple voices and surfaces even when it seems most innocently singular; he knew how we hear *between* notes, how a series of tones can create multiple surfaces. Bach's Prelude No. 1 in C Major, for example, is a simple composition, learned by most beginning piano students. It consists of a repeated pattern which is hardly melodic at all. The pattern is seemingly propelled out of its lowest tones, reaches a peak, and takes a short step back to repeat the movement upward. It is not a wave but a rocking.

We will come back to this pattern and its significance, but what will stay in the ear most forcefully will be those peak notes of the rocking. What we hear in this prelude are the peaks forming their own melody—a sort of halting, stuttering melody, on weak metrical moments, nestled in the repetition and clarity of the surrounding pattern.

What sense does this sort of melody make? We have already seen that melody has a structure, that it echoes itself, expands, breathes, returns, even—in tonal music—resolves. The more subtle the melody, the more nuances and demands and echoes it provides. And, put in simplest terms, resolution of a melody in tonal music means a return to ①. Bach's peculiar, off-beat melody begins in the first two measures by rising, from ③ to ④; it seems about to descend, as if beginning again, only to leap to ⑥. These are slightly disorienting movements, unsettling in their motion. We know from the surrounding rocking pattern how strong the sense of that unsung ① is; its harmonic shadow lies under the voice we hear. But that voice, hovering weakly on ⑥—then glancing off ⑤—is finally abandoned in measure 8. Immediately after that it is as if a different voice were beginning its motion. We are left seemingly suspended with the echo of that ⑥—and that opening ③—which never descend in an orderly fashion to ①. They really hang in the musical air: this can be heard; it can be felt. It is not until the penultimate measure of the entire prelude, when the repeated pattern has been broken and the piece is about to close, that we feel a melodic resolution, a step downward, ② closing finally on the desired ①. The piece requires feeling an arc of suspense that extends over its entirety. The melody should be felt, by the listener and the player, as a sort of surface whose continuity is unbroken. The shape of that surface is given by the laws of tonality.

This surface is really a line—there is only one dimension in which movement takes place. What gives music added resonance is that each point on that line is also a participant in other relations, intersecting and interacting with them. If we look closely at the patterns in this Prelude, for example, we see multiple lines fulfilling different functions. The lowest tones in each measure form their own sort of melodic lines, which in hearing can be even more pungent and powerful than the more easily heard soprano line.

But these melodic lines are contained in another sort of musical space that incorporates the tensions between tones: the space created by tonal harmony. The multiple lines in Bach's pattern, for example, are not only heard as lines. The melodic tones are also heard as parts of vertical units, chords, arpeggiated, spread out, sounding one note at a time. This arpeggiation is what gives the music its small-scale waves and pulse, underlying the melodic shapes; the chords create yet another surface defining the shape of the work.

In Western harmony such chords create a complex space. The traditional naming of chords by Roman numerals (like I and V and V_7) is actually a mapping of our sense of chordal tension and distance into a set of numbers, creating a formal structure like that of a mathematical system. It can resemble a topological space. There are rules for moving from one point to another on a harmonic path. There is even a sense of distance and neighborhood. And there is a quality of connection—attraction or repulsion. Some chords increase a sense of tension, others decrease it. There can be a sense of potential energy latent in a given harmony, comparable to the energy we say is latent in a weight when we lift it to a certain height: upon release, there can be a rush to a place of equilibrium.

First Prelude (BWV 846) from Bach's The Well-Tempered Clavier, *Vol. I.*

The chords carry some of the character of scalar pitches. The more important are the chord built on the fifth degree of the scale (V)—the dominant—and the chord built on the first (I)—the tonic. In the prelude the play between these harmonic poles appears throughout; Bach teases us with their reappearance and seeming resolution—in the third measure, and then in the seventh, the eighteenth, the twenty-fourth, and finally the thirty-first measures. In a sense, the spirited rocking pattern of each measure is replicated in the sort of repeated ebbing of harmonies in the piece. It is not simply a linear motion but a rising and falling, smaller patterns contained within larger ones. If we listen closely to this work when played by an artist who understands its various tensions, we might hear a larger pattern containing the smaller ones, reaching a grand structural point of suspense in the twenty-fourth measure—when the V chord makes itself fully felt. We will also hear something else four measures before the end of the work, something we may experience as a premonition of its closing; it consists of four measures of a chord that are effectively a I. This harmonic resolution coincides with the melodic resolution I have already described. Moreover, the bass line of the final cadence actually makes a thematic reference to the opening gesture of the fugue which is meant to follow the Prelude.

The study of tonal harmony—and, indeed, the performance of tonal music—involves learning to sense and interpret the nuances of such tensions and resolutions between chords. These relationships can become even more elaborate when the composer twists the frame of reference by changing the tonic. The composer can turn another chord into a I by a process called modulation. Modulations allow the creation of whole musical regions under the governance of different centers of gravity; it becomes possible, with modulation, not just to shift ground but to create large-scale relations between different grounds. Thus, we sense tensions not only within a key—that is, within a universe determined by a particular tonic—but also between that universe and another determined by a different key. A V chord may become a I chord; as it does, new relations are created between all tones.

In the Classical musical style that Charles Rosen described, these large-scale relations of key become central to the drama. In a Classical sonata we may play not only with a central harmony but with whole sections of music that create long-range tensions with other sections. Resolution in the Classical sonata, for example, might come not just with the return to the original I chord, or with a melodic descent to an original ①, but only when all the material heard in such a key as the dominant (V) is heard again in the tonic (I). The notion of resolution, of logical musical completion, applies not just to the small-scale detail but to the large-scale structure. Layers of relations extend tensions over time, as if we were passing over some strange spiraling surface. (And these remarkable effects do not even touch upon rhythmic relationships or instrumental timbre or tempo or dynamics, all of which combine to create a composition's inner life.)

We can understand these relationships a little more clearly if we "read" through one of Beethoven's piano sonatas. This reading is not a full-scale analysis; in fact, musicologists themselves are uncertain as to what a "full-scale" analysis would be. The greatest compositions are overdetermined; there are so many relations established and so many connections made in them that the problem with analysis is not where to begin but where to stop.

For example, consider the first movement of Beethoven's *Appassionata* Sonata. This work's dramatic impact is inarguably strong, which is why its spurious title has had such a long life. It begins with a figure that arpeggiates the tonic F minor chord. But the arpeggiation is given a thrusting rhythm that propels one forward; no stability is suggested. Each note is doubled by its compatriot two octaves away, so the sweep itself extends

Opening of Beethoven's Appassionata *Sonata*

through the heart of the keyboard, from a low F to a high F. It is an ambitious, spacious proclamation of the harmony that is also ominous because so softly played. The tonic tone, F, the ①, falls on accented beats and is held for the longest durations during this opening arpeggio, but there is a hint of ambiguity in the beginning and the end that doesn't let us feel that F is truly a tonic. The arpeggio ends as if poised to leap rather than prepared to rest.

That is because it lands at a point of greatest tension at the beginning of measure 3—the fifth degree of the scale, a C. We get a sense of having suddenly shifted focus or attention unexpectedly, an impression reinforced by the change in harmony, from tonic to dominant, at that point. It's as if we had come upon an unexpected vista. We hear a C suggestively hanging in the air, but we don't know how long it will last. When it moves it does something completely surprising, giving way to a trill on a tone that isn't contained in the F minor scale at all—a D♮. This pitch is foreign to the frame of reference established by the arpeggio; that alien condition is emphasized— *italicized*—by the trill. The trill then sinks back to a C, forming a three-note motif (C-D♮-C).

The motif's juxtaposition with the opening arpeggio is shocking. Grandeur is replaced by focus, sweep by stasis. Instead of an arpeggio we have, musically, almost the opposite—movement up to a neighboring tone. The tenor of the whole passage changes. The motif also seems ornamental, but it is also too unsettling to be ignored. Its close is on ⑤; its harmony a V. We are suspended in a tonal space. Its only connection to the preceding measures is a shared proportion—the ratio between the durations of the D and the C is the same as the ratio between the long and short notes in the preceding arpeggio—five to one.

The arpeggio, the thrusting rhythm, the neighboring tone, the relationship between the C and that D♮, the contrast between

major and minor modes—these are the ingredients of this work.
What is being posed, really, in this opening theme, is a sort of
musical problem in the relationship between these elements.
The theme is a statement of tension, of incompleteness, of
conflict, that contains the premises for the composition as a
whole. It is given still more importance by the extended pause
after the opening statement, as if that frame of silence were
required to make the internal oppositions stand out.

After that pause, the theme enters again, but in a completely
unsettling fashion; it is as if the whole musical surface had been
lifted and put on a different plane, bearing no relation to the
first. The entire theme (with a small change) is elevated a half
step. This is a small difference in pitch, but a great difference
in tonal topology: there is no close relationship between I and
II♭.

But the shift does have some relationship to the nature of
the theme. For in this second statement of the theme a D♭
appears wherever a C appeared previously. D♭ is even the first
note of this second statement; it follows the long pause. And
the ear hears it as a *response* to the C-D-C motion that disrupted
the first statement of the theme. Moreover, unlike the D, the
D♭ is actually the note we would expect to find in F minor; it
is the note we would have expected in the motif (it should have
been C-D♭-C). So its appearance in the second statement of the
theme is, in a peculiar sense, both a correction of the D♮ and
a reminder of its intrusion. The restatement of the theme focuses
our attention on the relationship between the C and the D.

The second statement of the theme can be thought of as a
mapping of the first, a transformation following the rule: "raise
by a half step." That transformation, because it raises all C's to
D♭'s, itself bears some relation to the melodic motion in the
second part of the theme, which contains a movement from C
to D♮. We now expect another movement, following the model
established by the first, namely a descent from the D♭ back to

the C—a sort of "correction" of the D♮ movement back to the C. Establishing this correction is actually one of the sonata's thematic concerns.

But first Beethoven elaborates on his idea. The second motif of the second statement (D♭-E♭-D♭) at once returns to the second motif of the first statement (C-D-C). It is as if all the relationships we have been hearing were foreshortened; our attention is drawn to the crux of the matter. Each C has become a D♭; now the D♭'s become C's. Unfortunately that D♮ still remains, sounding outside the frame, a rebuke to any promise of resolution. The essence of this thematic conflict is then abstracted and refined still further, as a figure is heard, murmuring in the bass, a figure that will mark each of the movement's climaxes. It is a compressed statement of the downward "corrective" movement I have described; it is unmistakable: D♭-D♭-D♭-C. This is hardly a subtle musical move. This bass figure reduces the contrasts between two statements of the theme into a motif. This motif leads to an even less subtle statement, a fortissimo outburst that comes to an end on a D-flat, which, in turn, descends to a C, on an insistent dominant harmony. The correction will eventually become an obsession.

So the second phrase of the theme (C-D-C), the emphatic bass figure (D♭-D♭-D♭-C), and the structural movement in harmony (I-II♭-I) and melody (C-D♭-C) between the two statements of the theme are all related: they are transformations or manifestations of a single concern. Indeed, the key of D♭ and the note of D♭ play a large structural role in the piece as a whole, marking harmonic climaxes and melodic turning points. In the tempestuous arpeggios and patter of the movement, all bass notes climb to reach that tone. But it is only in the final statement of the theme, in the coda (measures 251–256) that the second motif ends not uncertainly, but emphatically, and the movement is not C-D-C, as when it was first heard, but C-D♭-C, completing the gesture within the home key, sound-

ing this "corrected" proclamation in all registers of the keyboard.

My description of Beethoven's music hardly qualifies as a
finished musicological analysis. I have been relying on metaphors,
on rough suggestions that would have to be filled out with pages
of detail about harmony and voices, identifying the paragraphs
of meaning, the extended melodic arches that reach over whole
sections of the work. I have not named chords and their inver-
sions and talked at length about the use of the Neapolitan sixth
in the Classical period. But I am interested here in suggesting
something unmistakable in the experience of the music, something which may not be immediately apparent but which is
essential to grasping the music's purpose and direction. This
piece is *about* something; it begins with premises and explores
them. It leaves us understanding that theme more profoundly
than we did at the piece's beginning.

There is a particular aesthetic at work in this description
and in the sonata itself. Rosen called the Classical style a style
of reinterpretation. This is not a mere metaphor. There is hardly
a measure in the *Appassionata*'s opening movement that is not
the result of an analytical transmutation of the material we have
been discussing. This was typical of Beethoven's style. Rosen
demonstrates how the harmonic structure of the *Hammerklavier*
Sonata is based upon a melodic motif of a falling third in the
opening theme; a large-scale tension between B and B♭ also
derives from that material. "This is perhaps Beethoven's greatest
innovation in structure," Rosen writes. "The large modulations
are built from the same material as the smallest detail, and set
off in such a way that their kinship is immediately audible."

The sonata form—and Beethoven's style in particular—has
come to represent the apotheosis of an entire tradition in Western music, influencing even those who have dissented from it.
The music is narrative in style; it has a plot. The sonata of the
Classical period—indeed, the music of much of the nineteenth

century forming the heart of the concert repertory—is novelistic: however we "map" this music with analyses or programmatic glosses, it seems to require a story, something that changes over time, develops, transforms, and is then restored. The sonata takes musical events or ideas and subjects them to reflection, dissection, expansion, resolution. These ideas are developed and transformed, subjected to risks and dangers, until they return, in the recapitulation, all harmonic tensions resolved, all dissections undone. The sonata style reflects the literary and philosophical ideas of early Romanticism—presenting a journey in which the hero begins from a point of rest and proceeds using the powers of reflection and analysis, eventually returning to his origin, transformed and enlightened.

The power of this narrative impulse should not be underestimated. To a certain extent it dominated the European musical experience from Beethoven through the early years of the twentieth century. Outside this period, we may need other metaphors for describing musical process. In the Bach Prelude, for example, we are involved less with the unfolding of logical narrative than with a meditative musing on pattern and relation. For understanding Webern's musical structure we need a different vocabulary altogether. In twentieth-century serial music, the "premise" is not always a theme in the sense we have been discussing. Instead, the series presents something like a "repertoire" for the composition: suggestions of its possible gestures and relations. The series's intervals and their relationships help determine the music's long-range character. But we do not necessarily hear something being "worked out"; the effect can be more formal, and more static—more like a sculpture than a novel.

Still, whatever the style of the composition, music always involves an attention to relations between tones and rhythms— to musical spaces filled with lines of force and implication. In listening we attend to premises and arguments, to similarities and transformations, all of which extend contrapuntally over the length of a composition. That is how we hear melodies; it is how we hear the Bach Prelude; it is how we comprehend Beethoven's narrative. There have been attempts to catalog the various formal operations music exercises on itself. To a certain extent, this is what Schoenberg did when he declared that in his musical system a series of all twelve tones could be subjected to transposition (moved up or down), inversion (turned upside down within an octave), and retrogression (played backward); these are just formulas for transformation. Others have attempted to define musical transformations as a kind of grammar. These transformations and mappings seem to lead, as in mathematics, to a kind of knowledge.

Understanding music—playing it properly or hearing it properly—means, to a great extent, using the music to come to an understanding about it. We intuitively feel that music teaches itself. This is a remarkable phenomenon. For music seems a closed system that does not refer to anything outside itself. But what does this mean? To be sure, if we listen to something often enough we get to know it. We can hum the themes or know where the climaxes are, or call it to our mind's ear when there is no sound. To recall music in this way means that we also understand something of the order of the music, the sequence of its events, its motivations. We come to a sort of understanding about musical issues much the way a chess master will recall and interpret a game, or the way a mathematician will comprehend the development of an idea through a proof. This understanding comes first from learning to hear the way music creates distinction and articulation, in recognizing, for example, that the opening measures of Beethoven's *Appassionata* constitute a theme, or in hearing the soprano tones of Bach's figuration as a melody. This articulation is something learned, like the way we learn to hear sentences and words in a new language; units of musical meaning begin to become clear. These units may not even be objects in the ways we usually think of them—for the boundaries between musical units of meaning are generally ambiguous or malleable. Even a single phrase can have multiple layers, with melodic, rhythmic, and harmonic ideas intertwined. But the significance of these articulated elements comes from the connections we make upon listening, as we recognize similarities and differences among the objects we have discerned. Understanding music involves making mappings and appreciating transformations between musical objects.

These mappings may involve hearing repetition of particular musical details, which may lead to hearing others. We may even come to hear, with greater subtlety, the kinds of tensions that are being established and resolved. We hear more musical details as we listen, but we also hear more of the relations such details establish with one another. We begin to sense—at an unconscious level at the very least—the way musical thinking proceeds.

This may apply to music of the most varied styles and cultures. A style, after all, may do no more than define the nature of the articulations and transformations—the restrictions on one or the other, the expectations one has of either. But the existence of such abstract musical objects and mappings is what allows us even to learn a style, to follow, for example, the rhythmic variation and melodic play of an Indian raga even if we have never heard one before. It is also what allows us to comprehend many compositions written during the last thirty years by composers like Pierre Boulez, Milton Babbitt, and Elliott Carter, whose relationships and variations are built not upon tonal laws and relations but upon laws that may have been invented for a particular composition. Here the nature of the transformations may become quite personal, audible mainly to specialized audiences; still, the nature of listening remains the same.

• • •

WHAT SORT of knowledge does music offer? In mathematics, one form of knowledge concerns things familiar in daily life—numbers, for example, or spaces and their qualities. But we can also obtain knowledge about objects which have nothing to do with the world, mathematical objects we have posited or invented. We might come to understand something, for example, about the infinitesimal, or about a space's fundamental group, neither of which exists in any tangible way available to our senses or even to everyday cognition. These understandings may ultimately bring us back to the world, but our knowledge is, at least at first, strictly mathematical. We know their properties within the systems we have established, their relationships to what is already known.

Musical knowledge is, apparently, similar to that more abstract form of mathematical knowledge. There are no discoveries in music that illuminate our physical universe. Instead, through music, we learn more about musical objects—phrases, chords, gestures. We come to understand the implications of tensions and relations (between that C and D♮, for example, in the *Appassionata*, or between the harmonic and melodic rocking in the Bach Prelude). The deeper our understanding of a composition, the more knowledge we possess about the musical objects within it. We may even witness the transformation of one musical object to another—a sort of musical mapping in progress—thus learning about the process of musical thinking as well. Something like this happens when we hear the harmonies and voices change in the Bach Prelude or hear the pitch change in the two statements of Beethoven's theme. The transformation becomes the music's subject as much as the material itself.

Although it is absurd to take an evolutionary or developmental view of an art like music—as if Beethoven were an advance over Josquin, for example—there are ways in which it makes sense to talk about musical knowledge growing over time. We certainly know more about certain sounds than was known five centuries ago—that is, we understand how sounds once forbidden or rarely used can be used to construct compositions. We can hear new implications and create different musical systems. Schoenberg took this to reflect our expanded awareness of the more distant overtones of the vibrating string. Consonance and dissonance, he argued, are not opposites; their definition "simply depends on the growing ability of the analyzing ear to familiarize itself with the remote overtones, thereby expanding the conception of what is euphonious." Thus, the once dissonant interval of the third is now thoroughly consonant; the once forbidden tritone is used with as much impunity as the fifth in much twentieth-century music. And even when music is written today in the tonal language, we allow more relations between tones, and experience more combinations, than were once thought possible (or permissible). The boundaries of what is considered musical have moved.

There are two aspects to this development. First, the boundary of what is permitted and what is forbidden has shifted. During the sixteenth century, certain intervals in counterpoint would not just have made bad music; they would have made no music at all. During the nineteenth century, the works of Pierre Boulez would have been considered not just bewildering music, but complete gibberish. This boundary has become sufficiently malleable that John Cage gladly asserted that *anything* we hear can be called music. We have, in fact, learned to find more things musical—just as mathematical explorations have absorbed more of numerical experience into our understanding and have granted mathematical status to systems, such as those of non-Euclidean geometry, which would once have been dismissed as nonsensical.

But if there is one thing the various musical experiments of the twentieth century have shown, it is that music requires the forbidden to create its order. In the Beethoven sonata, the premises permit certain kinds of harmonies and "forbid" others. The presence of an invisible boundary provides drama: this is surely the significance of that D♮, protruding in the opening theme. It is a visitation from someplace outside the world of the arpeggio.

It may be that the entire concept of musical dissonance should be understood in this way—as a musical rendering of the challenge of nonmusic. It is the introduction of noise into order, the threatened dissolution of space and field and surface into mere events, isolated points; it is, in short, the specter of discontinuity. The operatic version of this threat is the gypsy in *Carmen*—or the courtesan in *La Traviata*—representing a force that lies outside the social order, that threatens it, but whose presence gives the drama its drive, and gives the order itself its reason for existence. The mystery of tonal music is that the relationship between two distinct notes with the greatest consonance—the interval of the fifth—also becomes the relationship of greatest challenge, the interval with greatest potential energy, the gap which music will try to close, a model for all other gaps, forbidden and dissonant.

Musical knowledge is thus something mysterious indeed. Music, to create an order, always involves constraints, limitations, and regulation of sound. But within that constraint, there is also an urge for testing and trying those very limits, maneuvering around the borders of the acceptable. Such maneuvering is Western music's drama, heard in dissonance and euphony, modulation and elaboration.

The growing freedom of the dissonance—the expansion of the territory—meant that the boundaries of music had to be set more and more within the music itself, rather than from surrounding notions of aesthetic form or social function. Each work has had to create its own laws. Music thus became more abstract in the twentieth century, as previous boundaries were breached.

TO RECAPITULATE: the knowledge gained in music is knowledge about musical objects—tones and intervals and phrases and harmonies and rhythms. It is also knowledge about how these objects are organized and associated. We learn to recognize musical objects in ever larger groupings, then learn to recognize when these groupings are transformed: what changes and what remains the same. We learn about musical mappings. This is how musical variations work, as I have suggested. We are given a theme; the variation is a mapping of the theme into musical space which preserves certain aspects of musical continuity. The mapping might be an instruction to "change the rhythm," or "shift from major to minor harmony," or "ornament the melodic line," or it might be "play freely but keep one element constant," as in Bach's *Goldberg Variations*, where the bass line is fundamentally unchanged throughout the most fantastically brilliant musical alchemy.

The variation is one special example of the kind of musical thinking that goes on in any composition. Music shapes time because it marks it with the experience of change and transformation. When we talked about space in mathematical thinking, we saw extraordinarily abstract concepts developed to explain and display aspects of physical space; we needed notions of smoothness and neighborhood, ways to characterize different organizations of space, and we soon had to leave ordinary experience behind as abstract systems were developed bearing little obvious relation to the physical world. Musical compositions are organizations of time, which, unlike space, is given its shape by the events occurring within it. Music commands time; it determines whether time seems to rush forward or hang suspended;

it determines when time is to become knotted, turning back on itself. These surfaces are dizzying because a single point in musical time can be host to a contrapuntal assemblage of events. This temporal space is so laden with crossroads that we can choose, in our listening, any number of paths through it.

A composition is an exploration of musical "space," then, which creates its own topology. It establishes which events are continuous, which discrete; it creates connections between small events and large (between the C-D-C of the opening phrase of the *Appassionata* and later gestures on a larger scale). The composer determines what paths may be taken through a musical space and where they will lead. We can even categorize these spaces by their forms and functions. And the entire work is consumed by process. Structurally, for example, the development section of a sonata is a series of mappings and transformations of the sonata's opening material; its recapitulation is also a mapping, which is not merely a repetition, for in music, no repetition is ever identical; our comprehension of it is changed by our passage through the work itself.

The processes within that abstract world have become the subject of increasingly close examination. Never have composers and analysts so explicitly examined the theoretical processes of music as during the last fifty years. Changes in musical style and language during the twentieth century apparently focused attention on issues which, for a long time, could be just taken for granted. It was surely the impending dissolution of tonality, for example, that led to Heinrich Schenker's profound and influential examinations of music at the end of the nineteenth century. He formulated an understanding of tonal music that incorporates nearly everything we have described so far: the sense of long line, of suspensions and transformations, of multiple voices taking different priorities. He came to regard all tonal pieces as elaborations and prolongations of a basic melodic line descending by steps to ① and all harmonies as elaborations and prolongations of a basic harmonic progression—I-V-I. He literally graphed compositions, showing which tones were structural and which ornamental at various perceptual levels or "layers." In Schenker's view, the very concept of modulation—of a shift into another tonal region—is unimportant; it is filigree on a space more fundamentally defined. Schenker talked about the foreground and background of a composition, its basic line, its fundamental movement. The implication of his analysis was that not all notes of a composition are of equal significance. Some are ornament, some are structural; some are filigree, others are indispensable. One harmony may be the syntectic crux of a composition, another a more slight fancy. These concepts are not inventions of analysis but the sorts of required distinctions made by the composer in constructing the work and any performer attempting to make sense of it. The process of making musical sense means making functional distinctions between tones and phrases and harmonies, perceiving that some are directly related and others are not, that some are transformations of others, and that some are punctuation. The composition must be decomposed, quite literally, so that the processes of construction become evident in sound. The structure must be laid bare.

There has been great controversy over Schenker's theories, mostly among analysts who believe that his vision of music is incomplete. But Schenker's attention to basic structures of tonal music and their transformations helps us understand all music. By the time tonal musical geometry had become just one system among many, the notions of musical transformations and structures had become as fundamental as the structures mathematics began to create and discover in its own activities. During the last few decades, the entire process of composition and musical thinking has become the subject of more extensive abstraction— in the critical work and music of composers such as Milton Babbitt and George Perle and in the analytical theories of Allen Forte. The movement has been toward generalizing our under-

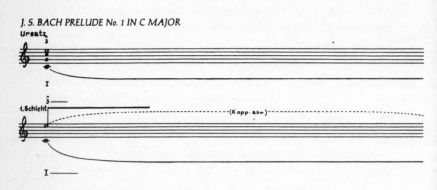

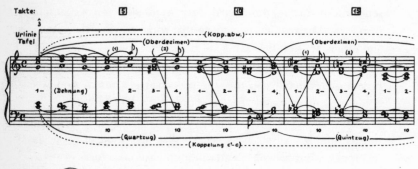

Heinrich Schenker's graphic analysis of Bach's first prelude from The Well-Tempered Clavier, *Vol. I*

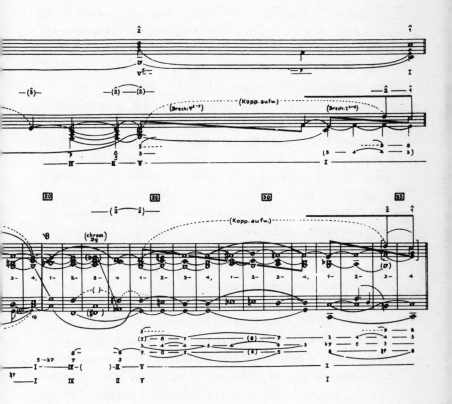

standing of music itself, even as a younger generation of composers has been either embracing neo-Brahmsian romance, or
attempting to discard any pretense of stylistic or logical consistency.

David Lewin, a theorist with mathematical training, has
mathematicized musical concepts in a remarkable recent book,
Generalized Musical Intervals and Transformations; he makes explicitly
mathematical some of the images I have been using. "In concep-
tualizing a particular musical space," he writes, "we often con-
ceptualize a family of directed measurements, distances, or
motions of some sort." He begins, that is, with a general notion
of a sort of musical topology—a musical space with elements
s and t, with a measurement of a distance from s to t. Depending
on our notions of what these elements are, and what distance
consists of, we may get vastly different systems of music. S and
t may be pitches in a twelve-tone musical space, in which case
the distance between them might be the number of semitones
between them. S and t may also be elements of a musical space
consisting of points in time with a distance of one temporal
unit; the generalized interval between s and t, then, would be
the number of temporal units by which t is later than s. Lewin
also suggests different spaces based on timbres, or rhythms,
durations, and relations between harmonies. He then argues that
all these examples—which turn out to be useful in the description of musical processes within a composition—share a certain
mathematical structure. They are groups—just as a topological
space is said to be made up of fundamental groups. They are
organized structures of musical elements that have certain kinds
of relations to one another.

What Lewin does with this concept—and with others of
still greater generality—is make more precise the whole notion
of transformation and variation. The abstraction he draws from
the examples presented to him seems almost to "lift" out essential relations: as in mathematics, the result is an articulation of

structure. The identification of a "group structure" in the ways we organize musical material helps explain some of the aesthetic requirements we place on music as well, our sense that music does create a closed world of relations, that it defines ways of moving and changing, that these are highly regulated.

Consider, for example, something as elementary as the notes on the piano's keyboard. This musical space is constructed with a system of musical intervals which relies on the notion of congruence: if an interval is the number of semitones between two musical elements, we consider two intervals congruent or equivalent if they differ by a multiple of twelve. Thus, a fifth, which is the distance from C to G, contains seven semitones; it is equivalent to a leap from C to the next highest G, which encompasses nineteen semitones. So linear space is wrapped around itself into a cylinder of rising pitches with all octave differences eliminated, like the movement on a face of a clock. There are only twelve classes that matter. Lewin specifies the sorts of transformations that might work in such a space—from such familiar ones as transposition and inversion to more exotic manipulations of timing and timbre. These formal definitions of musical spaces focus our attention on what sorts of processes take place within music.

This technique may even help us to hear something in the music; once the analytical tools are applied to a work, relations are revealed to the ear as well as the mind. But Lewin's arguments, as exquisitely structured as they are, are still preliminary; they just begin to address the ways musical form is constructed and musical argument takes place. This is not surprising. Music is not mathematical; it simply shares some of mathematics's techniques and manners of thought. The attempts by Lewin and other theorists like Robert Morris to specify the techniques of tonal music in a rigorous fashion have grown out of much more rigorous techniques used to organize music in the twentieth century.

The image of an organism has often been used to show how music grows out of principles encoded within itself, how its large-scale structures and small details can be intimately related. But calling music organic treats it as something that develops rather than as something created; it treats music as a process, determined in its detail and form, rather than as something more closely related to the kind of mathematical activity we have discussed. If music is organic it is because it is something else: a representation of thinking. Music's musculature is constructed out of the sinews of argument—even when the result is as seemingly mindless as the latest ad jingle. The arguments and relations may vary from style to style. But the character of musical thought is unmistakable: it even reflects upon itself, looking back, recalling elements and ideas from the past, transforming them, deriving new ways of hearing from old. The Classical sonata form characterized by Rosen may even be properly and persuasively regarded as a musical model of Hegelian thought.

The theorist David Epstein presents a subtle view of compositional unity in his book *Beyond Orpheus*, showing that the "premises" a composition broods upon can extend to nuances of timbre and tone, duration and character. He suggests a point of view "reminiscent of Aristotle's idea that the form of the statue resides a priori within the material, awaiting its revelation by the sculptor. This seems almost a paradigm of compositional thinking: a musical idea, once conceived, has a variety of possible manifestations; it is the composer's obligation to realize these and to give them concrete form."

The creation of concrete form in music involves seeing the microcosm within the macrocosm, using small relations as models for large ones, transforming music according to the rules it establishes, through the demands and commands of a style and through the particular domain of a given work. To carry forward our analogy with mathematical explorations, a composition begins out of its premises and proceeds to "prove" itself. When correctly understood, this is the way a composition is sensed— as an inarguable fact, which could hardly be different. Great compositions create their own form of necessity, the binding coming not from logic but from the unfolding of ideas with all the variation and style and imagination we find in sophisticated mathematical proofs. Much of the profound music of our culture is "intellectual" only in that it takes the powers of intellect seriously in the texture of music itself.

THE PATHS we have taken have led through difficult terrain, but we have already begun to get some sense of the vistas that lie ahead. Mathematics and music share certain techniques, uses of forms, ways of thinking. Beethoven's exploration of a single motif required subtle methods of abstraction and juxtaposition; Bach in his Prelude created echoes of small-scale detail in large-scale structures. A mathematician faced with similar abstract forms would have asked similar questions and, perhaps, given similar answers. There are similarities that exist because mathematics and music take on projects of abstract exploration and are guided in their explorations by processes of analogy and transformation. There is no strict formalism, of course, a system which, like the one dreamed by David Hilbert in mathematics, would somehow generate itself. But the images, the mappings I have been suggesting do not have to be—to use a mathematical term—isomorphisms. There is no identity suggested between music and math. There is, however, a series of relations that are more than merely suggestive. We see some of those relations now on a purely formal level.

Were our ambitions less demanding, all that would be needed, it might seem, would be to expand the arguments already made, filling them with more rigorous analysis and greater detail. The notion of dissonance in Western music would have to be examined just as we examined the concept of the irrational in mathematics; the analytical scholarship would have to be surveyed to get a clearer idea of how argument in music is constructed. We might also explore the notion of compulsion in mathematical proof and show how style in mathematics is linked to the function envisioned for mathematical ideas just as musical styles have been linked to function. After this additional labor, expanding our perspective without changing it, we might not be faced with the sunrise from the top of Mt. Snowdon, but we surely would have a scene of extraordinary intellectual grandeur.

These considerations though would still be *merely* formal, demonstrating the ways in which elements of music and mathematics arise from other elements, the ways in which the rules and gestures of compositions and proofs relate to one another. We have begun to appreciate musical and mathematical thinking, but we have not yet begun to understand the way they "feel," the ways they grant to us some vision or sense of things, why they attract us, and what revelations they offer.

Where do music's beauty and power come from? If music is as formal as I have been arguing, how do we return to the aspects with which our exploration began and see it as an earthbound, timebound creation? The same questions can be asked of mathematics. I have suggested, of course, that mathematics always bears some relation to the physical universe. But in what way do its formal constructions reveal the universe? And what relation does its beauty have to what seems to be true?

For the answers we must partly turn back, surveying the paths already taken from a broader perspective. We begin again, this time not with questions about logic and reason and method but with questions about beauty.

IV

THEME AND VARIATIONS:
THE PURSUIT OF BEAUTY

The mathematician is only complete insofar as he feels within himself the
beauty of the true.

JOHANN WOLFGANG VON GOETHE

THE FORMAL CONNECTIONS BETWEEN MUSIC AND MATHE-
matics, the ways in which each art is also a science, may
now be more evident, but we must come closer to their inner
lives. We must begin to see them from within, to understand
not just how their processes work but how they act on their
devotees and what promises they offer.

The inner life is not easily reached from outer appearance.
No purely objective analysis of *The Marriage of Figaro* can match
the experience of Mozart's music; no description of formal relations can account for the almost transcendental qualities of the
music of Bach. Robert Schumann wrote upon hearing Hector
Berlioz's *Symphonie Fantastique*: "Berlioz, who studied medicine
in his youth, would hardly have dissected the head of a beautiful
corpse more reluctantly than I dissect his first movement. And
has my dissection achieved anything useful for my readers?" An
analysis presents only a single narrative about a piece that can
tell many tales at once; analysis has strict limits and may even

miss music's essential power. There is a vast gap between what we say and what we feel about music.

In mathematics the inadequacies of mere analysis are still more clear; as we have already seen, a three-line proof represents years of work and consideration. But at least in mathematics we can begin to get some sense of the inner life of the art through the accounts of its creators. Henri Poincaré, in a lecture about mathematical creation, told of his labors attempting to show the existence of a group of mathematical functions called Fuchsian. Here are his oft-cited words:

> Just at this time, I left Caen, where I was living, to go on a geologic excursion under the auspices of the School of Mines. The incidents of the travel made me forget my mathematical work. Having reached Coutances, we entered an omnibus to go some place or other. At the moment when I put my foot on the step, the idea came to me, without anything in my former thoughts seeming to have paved the way for it, that the transformations I had used to define the Fuchsian functions were identical with those of non-Euclidean geometry. I did not verify the idea; I should not have had time, as, upon taking my seat in the omnibus, I went on with a conversation already commenced, but I felt a perfect certainty. On my return to Caen, for conscience's sake, I verified the result at my leisure.

This tale of illumination—earlier eras might have called it revelation—is all the more famous because it is not untypical. Carl Friedrich Gauss gave a similar account of a theorem he had been unable to prove for years: "Finally, two days ago, I succeeded, not on account of my painful efforts, but by the grace of God. Like a sudden flash of lightning, the riddle happened to be solved. I myself cannot say what was the conducting thread which connected what I previously know with what made my success possible."

Jacques Hadamard, in an influential little book, *The Psychology of Invention in the Mathematical Field*, compared the sense mathematicians have of seeing the answer to a problem instantly and in its entirety with Mozart's letter about seeing his symphonies whole in his mind before setting them down on paper. The "seeing" is not literal; it is not pictorial. It is a sense instead of grasping something at once, correctly and completely.

Poincaré suggested that the vision reveals a kind of choice. The rules for such choice, he went on, "are extremely fine and delicate. It is almost impossible to state them precisely; they are felt rather than formulated." He connected insight into mathematics to "emotional sensibility":

> It may be surprising to see emotional sensibility invoked apropos of mathematical demonstrations which, it would seem, can interest only the intellect. This would be to forget the feeling of mathematical beauty, of the harmony of numbers and forms, of geometric elegance. This is a true aesthetic feeling that all real mathematicians know, and surely it belongs to emotional sensibility.

This feeling of rightness and wholeness has something to do with our ideas of beauty. And while beauty is an unquestioned aspect of music, we may learn more by following Poincaré's lead into mathematics.

WE MUST RECALL, first, that there is something known as mathematical style, which we briefly explored in "Partita," the second chapter. We can speak, for instance, of the "rhythm" of a proof—the ways ideas are introduced, the kinds of punctuation that are used, the imitations of generic models—archetypes of argumentation. There are moments of drama, surprise, even compositional styles in mathematical proof. Here is the great nineteenth-century physicist Ludwig Boltzmann on his still greater colleague J. Clerk Maxwell:

> Even as a musician can recognize his Mozart, Beethoven, or Schubert after hearing the first few bars, so can a mathematician recognize his Cauchy, Gauss, Jacobi, Helmholtz, or Kirchhoff after the first few pages. The French writers reveal themselves by their extreme formal elegance, while the English, especially Maxwell, by their dramatic sense. Who, for example, is not familiar with Maxwell's memoirs on his dynamical theory of gases? . . . The variations of the velocities are, at first, developed majestically; then from one side enter the equations of state; and from the other side, the equations of motion in a central field. Ever higher soars the chaos of formulae. Suddenly, we hear, as from kettle drums, the four beats, "put $n = 5$." The evil spirit V (the relative velocity of the two molecules) vanishes; and, even as in music, a hitherto dominating figure in the bass is suddenly silenced, that which had seemed insuperable has been overcome as if by a stroke of magic . . . This is not the time to ask why this or that substitution. If you are not swept along with the development, lay aside the paper. Maxwell does not write program music with explanatory notes . . . One result after another follows in quick succession till at last, as the unexpected climax, we arrive at the conditions for thermal equilibrium together with the expressions for the transport coefficients. The curtain then falls!

It is not necessary even to understand the details of Maxwell's argument to get a sense of its spirit from this description. It is also no accident that Boltzmann should have used the imagery of Romantic symphonic music to give an account of Maxwell's style of argumentation. Maxwell was the man, after all, who, during the same years these symphonies were being written, had composed a brilliant series of equations connecting the phenomena of electricity and magnetism with the concept of invisible "waves" in space—insights which were still resonant. There is a Romantic spirit evident behind those equations and behind Maxwell's other work; it would not be difficult to imagine accompanying sounds of kettledrums, chaos, climaxes, and a final harmonic equilibrium.

It requires intimate familiarity with the gestures of mathematical reasoning to be sensitive to such aspects of style, which are present in all mathematical work. But style is not just a casual concern of mathematicians (or mathematical physicists). What we have been calling style is actually an aspect of the emotional sensibility of mathematics, part of its inner life. Beauty is one of mathematics's goals.

That goal can even take priority over others. The mathematician Hermann Weyl wrote, "My work always tried to unite the true with the beautiful; but when I had to choose one or the other, I usually chose the beautiful." G. H. Hardy, England's leading mathematician in the first half of the twentieth century, who wrote one of the most beautiful memoirs by a mathematician, asserted more dogmatically that "there is no permanent place in the world for ugly mathematics." And Bertrand Russell, who, with Alfred North Whitehead, devoted not a little energy to the attempt to systematize all of arithmetic using the symbols and syllogisms of mathematical logic, affirmed, "Mathematics, rightly viewed, possesses not only truth, but supreme beauty— a beauty cold and austere, like that of sculpture . . . The true spirit of delight, the exaltation, the sense of being more than Man, which is the touchstone of the highest excellence, is to be found in mathematics as surely as in poetry."

Anyone who has ever listened to a lecture by a physicist or mathematician whose understanding of the subject is profound as well as thorough can also provide accounts of such aesthetic appreciation. One student of the nineteenth-century mathematician Charles Hermite wrote, "Those who have had the good fortune to be students of the great mathematician cannot forget the almost religious accent of his teaching, the shudder of beauty or mystery that he sent through his audience, at some admirable discovery or before the unknown."

When I studied advanced mathematical analysis with Shizuo Kakutani and followed his extraordinarily compact arguments and elegant solutions, or heard Abraham Robinson lecture almost casually about his advances in mathematical logic, it was not the drama of the speaker's voice that caught me: in neither case was the lecture a "performance" like those familiar to students of the humanities. The lecture was an unfolding, taking place so subtly that, by the time the classes were completed, there was a sense that one had been led into regions one had not known existed. There were times when it seemed as if I had understood the lecture, as if all the details were in place, only to find out, during the following week, that the more I thought about the lectures the more intricate and subtle I realized the arguments were. In other words, I had the same sense one has when confronting an artwork whose powers seem more mysterious than those of simple propositions and whose influence can be longer lasting than any other experience of our senses. Even if we have never felt a shudder of beauty or mystery, or associated it in any way with mathematics, we can attempt to comprehend it.

The mathematician Rózsa Péter gives an example of thinking that may help us understand these issues better. Here is the riddle Péter poses in an article in *The Mathematical Intelligencer*:

We have two identical glasses; in one, we pour wine, in the other, water, to the same height (not quite full). Then we take a spoonful of wine from the first glass, put it into the glass with the water, and then mix it. Next we put a spoonful of this mixture into the wine of the first glass. Thus, as the end result, some wine goes into the water and some water into the wine. Which is more: the pure wine that went into the water, or the pure water that went into the wine?

Most of us, hearing this riddle, would immediately reply: more wine must go into the water, since the spoonful of liquid going into the water contains pure wine, but the spoonful going into the wine contains a mixture of wine and water. But there is another factor to consider: the spoonful returning liquid to the wine comes from a mixture that contains *more* than a full glass of liquid, and it is being put into glass that is *less* than full, so the mixtures are a little bit more complicated.

Our first temptation upon grasping the complexity of this apparently simple question is to work out a calculation. What if, instead of a teaspoon, we were to use a half cup measure and large glasses each containing a cup of liquid? Then half a cup of wine would be removed, leaving a half cup in the glass. When that half cup is put into the full cup of water, we get a cup and a half of liquid, which is one third wine. When we then take a half cup of that mixture to put in with the pure wine, that half cup will contain a sixth of a cup of wine and a third of a cup of water. But since we are removing just a sixth of a cup of wine from the mixture (having placed a half cup there) we are leaving a third of a cup of wine in the water.

This solution does not seem to depend on the size of container we use to remove liquid, so we may have actually "proven" that each container will have equal amounts of the other's liquid. But this is not a very compelling answer. It still violates our "common sense" and requires too much calculation. It is what mathematicians call a brute force solution. Péter even provides a similar solution she thinks physicists are likely to work out: if a spoonful of the mixture contains a quarter of a spoonful of wine, then from the spoonful of wine that was taken out, we are putting back a quarter of a spoon. Hence three quarters of a spoon of wine remain in the mixture, while three quarters of a spoon of water is placed in the wine. This is more elegant than the solution we would have been driven to, but to Péter it shows a prejudice in favor of "calculation."

Here is her solution:

Exactly as much wine goes into the water as water goes into the wine. For let us consider, say, the wine glass at the beginning and at the end, ignoring what happened in between. The liquid in it stands exactly as high at the end as at the beginning. The difference is just that it has lost some pure wine (which is now in the water glass) and gained some pure water. If its loss were greater or less than its gain, then the liquid would have to stand lower or higher than at the beginning. Thus its loss—the wine that went into the water—is exactly as much as its gain: the water it received.

I suspect that this answer will need to be read several times before its force becomes clear. It involves no calculation, no hypothetical proportion; it seems so simple that it appears at first to be incorrect. Yet it does, as Péter asserts, get at the essential element in the problem, undistracted by other considerations. It simply compares two states of things, and argues backward. It demonstrates, she argues, the way mathematicians think.

It also gives some hint of what an "elegant" argument is in mathematics. Indeed, there is something beautiful about this way of thinking, so surprising, compelling, and refined are its strokes. But it is not a beauty of nature we are recognizing: we don't come away from this problem with greater appreciation for glasses of wine and water; we come away with a sense of the texture of thought itself. The beauty is in the simplicity of the reasoning when faced with the complexity of the expected calculations.

Here is a more explicitly mathematical example. Carl Friedrich Gauss was one of the great prodigies of mathematics. The story is told that in 1780, at the age of three, before having been taught any arithmetic, he had discovered enough on his own to correct his father's sums. When he was ten his school class was given a long addition problem. They were asked to add the numbers from 1 to 100 (or some similar series). Gauss thought a moment, then wrote down the answer: 5,050. The trick of such a calculation had been known, of course, but not to the young Gauss. Nevertheless, he recognized that if you take the first and last numbers in the series (1 and 100) and add them, you get 101. If you take the second and second to last, you also get 101, and so forth until you add 50 to 51. There were fifty such pairs of numbers. Hence the answer was 50 × 101. There is something more wonderful about the method. For it uses a property shared by all series of numbers which differ by an unvarying amount: the numbers 8, 15, 22, 29, . . . 708 could be added in a similar way.

This is a "trivial" problem, as mathematicians would say, but the solution is, nevertheless, beautiful. It would be quite possible simply to add the numbers, one by one. Instead, the method recognizes a pattern, then uses that pattern to understand something about the symmetry of a series. The beauty of the method is also due, in part, to the size of the problem proposed (add a hundred numbers, a thousand numbers, a million numbers) and the simplicity and ease of the method presented (all one needs is the first and last number and the number of numbers involved). The contrast between the enormousness of task and the simplicity of method creates a sense of great release; it is like hearing a joke whose punch line confounds expectations and turns expectation into laughter. It also is sufficiently abstract and general to provide something quite powerful: a way of thinking.

In more serious examples, the elegance is more profound. It involves changing our conception of an object by clarifying something essential about it. This can affect our understanding of daily experience, the way, for example, understanding the nature of mathematical continuity can reveal the way we experience ordinary continuity. But for our purposes the beauty is more transparent the further the argument is removed from intuition.

For example, one mathematical creation that challenges our intuitions of continuity is called the Cantor set. Here is how it is constructed:

Take a line one unit in length.

Remove the middle third, leaving

---   ---

Then remove the middle third of each of these two thirds, always keeping the end points of the lines in the set.

-- --   -- --   -- --   -- --

- -   - -   - -   - -

If we do this an infinite number of times—always removing the middle third of the remaining line segments—we have a set of points we cannot even draw, much less see. But we can add the line intervals that have been removed. First we removed a line with a length of $\frac{1}{3}$, then segments with a length of $\frac{2}{9}$, then $\frac{4}{27}$, and so on. The total length of these lines, after an infinite number of such additions, is 1—which means the total "length" of the line is extracted, leaving us with a set of points with 0 length. Yet the number of points remaining in this lengthless set is infinite.

This is an important illustration, for it is a mathematical construction that forces us to refine our concepts of length and continuity. Such constructions have had a distinguished place in the history of mathematics, at times answering fundamental questions. For example, one of the major discoveries of the nineteenth century, by Georg Riemann and Karl Weierstrass, was of a function—a curve—that is everywhere smooth but at no point has a specific direction; we can move easily from one point to the next (there are no gaps), but another aspect of "smoothness" is missing, the sense that we are going *someplace* as we move along it. Such mathematical creations (which often become objects of their own fields of study) have an eerie beauty. They display an intriguing balance between the inventive and the concrete. Often, there is in their creation a crucial moment on which the overall argument turns.

In the problem with the wine and the water, for example, the moment is when Péter discards all the intermediate steps and considers just two factors—the state of the wine glass at the beginning and the state of the wine glass at the end. In the addition problem it is when the arithmetician discards the ordinary sequence of numbers being presented and, through the reordering, finds that it is sufficient to add only two numbers— the first and the last. In the Cantor set it is the recognition that we are removing the entire length of the line when we create the set. In all these examples, we are being presented with an argument that seems thoroughly conventional. Yet the familiar suddenly becomes strange. Like a walker who is so intent on moving his feet that he forgets to look around, the mathematician is transported and transformed. The apprehension of the beautiful has something to do with this experience.

Many of the proofs we explored in the second chapter have the same character. Cantor's diagonal proof, Euclid's prime proof, the concepts developed in topology all lead us into unexpected regions given the nature of the problems posed; and all bring us to resolutions with an extraordinary efficiency and elegance. The nineteenth-century mathematician Lord Rayleigh wrote, "Some proofs command assent. Others woo and charm the intellect. They evoke delight and an overpowering desire to say 'Amen, Amen.' "

The beauty in these examples of reasoning is also connected to the real world out of which mathematics evolves; each reveals something about the general, not just about the particular; and each has something to do with truth. The beauty of the true is an inescapable aspect of the inner life of music and mathematics. However abstract mathematical reasoning may seem, through it we can begin to see the world as it really is, to discover hidden relationships and connections between events that otherwise seem disconnected. Gauss led us to understand not just one but all regular arithmetic series; the Cantor set led us to see that there is no necessary connection between our ideas of length and the accumulation of points. After some contemplation the beauty of the mathematics begins to reveal a more profound beauty in the world itself.

CONSIDER as an example of mathematics's intricate dance between the real and the created, between interpretation and invention, the number π. Recall: it is equal to the ratio between a circle's circumference and its diameter. No matter what size circle—no matter what length the diameter—this number is constant. It is considered equal to 3 in the Bible ("And he made a molten sea, ten cubits from the one brim to the other; it was round all about, and his height was five cubits; and a line of thirty cubits did compass it round about"); the Babylonians found it to be $25/8$, and the Egyptians identified it as $4(8/9)^2$. In 1767 π was proven to be not rational: it cannot be expressed as the ratio of two integers at all. It then became a bit of an international contest to see to how many digits of accuracy π could be calculated (for, as we recall, an irrational number can only be expressed as an infinite nonrepeating decimal: e.g.,

$\pi = 3.14159265358979\ldots$). Mnemonic poems were developed to aid in remembrance of such an eccentric collection of digits; the number of letters in each word would provide the decimal number for π. Here is a French example:

Que(3) j(1)'aime(4) à(1) faire(5) apprendre un nombre utile aux sages!
Immortel Archiméde, artiste ingènieur,
Qui de ton jugement peut priser la valeur?
Pour moi, ton problème eut de pareils avantages.

It turns out that there is no discernible pattern in the digits of π—despite the promise latent in the French verse. It also turns out that π is even more than irrational: it is *transcendental*. That is, there is no ordinary algebraic equation that can be "solved" for π the way $x^2 = 2$ can be solved for the irrational number $\sqrt{2}$. So π is one of those numbers which are calculated with extensive precision by exotic computer programs but cannot, of course, ever be calculated exactly. This would seem a terribly awkward position for a number of such fundamental importance to the beautiful and regular circle.

Yet centuries of mathematical exploration have, in fact, yielded formulations for π that are startling in their elegance. In 1665, for example, John Wallis came up with this formulation:

$$\pi = 2\,\frac{2 \times 2 \times 4 \times 4 \times 6 \times 6 \ldots}{1 \times 3 \times 3 \times 5 \times 5 \times 7 \ldots}$$

The philosopher Leibniz found this expression:

$$\frac{\pi}{4} = 1 - \frac{1}{3} + \frac{1}{5} - \frac{1}{7} + \ldots$$

And Euler, the mathematician who bequeathed to us nearly all of our notation and many of our techniques, found the sum of an infinite series which had eluded the grasp of mathematicians for centuries—the series of inverse squares:

$$\frac{1}{1^2} + \frac{1}{2^2} + \frac{1}{3^2} + \cdots$$

The sum?

$$\frac{\pi^2}{6}$$

It isn't necessary for us to explore how such formulations were derived. It is enough to see that something extraordinary is revealed in them; the seemingly random digits of our decimal expansion of π have given way to expressions of elegant simplicity: organizations of odd and even numbers, sums of simple fractions, exotic combinations of familiar objects. In these forms, π is revealed as the unusual number it is. It is as if we suddenly looked at the world through a glass prism and saw a pattern that was not evident before, or as if we were hearing a composition we could not understand until we realized that we were listening for entirely the wrong things.

But Euler went further. There are two other numbers as important to mathematics as π, namely e—the number that has a subtle relationship to logarithms—and i—the "imaginary number" equal to $\sqrt{-1}$. Neither of these numbers need concern us right now (though the number i is a wonderful example of a mathematical invention—an ideal object—that turns out to be extraordinarily useful among real objects). Accept for the moment the unusual character of both these numbers; then consider the formula Euler presented:

$$e^{\pi i} + 1 = 0$$

Here is a result of unexpected simplicity, connecting mathematical objects that seemed to have nothing to do with each other. Even without understanding such a formula, it is as if some glimpse has been granted to us of another world which was previously hidden. This is not just the beauty of mathematical invention. It may have taken Euler's genius to have so easily explored matters others passed by, but it seems clear that something has been revealed of these objects' inner nature, something philosophers might call the in-itself, the timeless, abstract Forms mathematicians have always claimed as their realm, and upon which mathematics models itself. Here the experience of order, connection, and pattern is found in numbers that seem to be beyond comprehension; there is a harmony established in which all elements are shown to be in hidden but profound connection.

This is the beauty of the True. It is a beauty that is not necessarily obvious. The self-taught brilliance of Ramanujan, for example, was more eccentric than Euler's, his formulas beyond easy comprehension. The mathematician G. N. Watson said about one such formula, "I would express my own attitude . . . by saying that such a formula as

$$\int_0^\infty e^{-3\pi x^2} \frac{\sinh \pi x}{\sinh 3\pi x} \, dx =$$

$$\frac{1}{e^{2\pi/3}\sqrt{3}} \sum_{n=0}^\infty e^{-2n(n+1)\pi} (1 + e^{-\pi})^{-2} (1 + e^{-3\pi})^{-2} \ldots$$

$$(1 + e^{-(2n+1)\pi})^{-2}$$

gives me a thrill which is indistinguishable from the thrill which I feel when I enter the Sagrestia Nuova of Capelle Medicee and see before me the austere beauty of *Day, Night, Evening,* and *Dawn* which Michelangelo has set over the tombs of Guilio de' Medici and Lorenzo de' Medici." Perhaps so, but this formula, even if we don't understand its significance, is more baroque and more full of ornamentation than anything Euler created. Here we are far more conscious of the creator's inventiveness, while Euler's formula suggests instead the genius of a discoverer. Both help reveal something objective, seemingly beyond creation and unaffected by discovery. Both somehow justify the aesthetic claims made by mathematicians. Poincaré wrote, "The Scientist does not study nature because it is useful to do so. He studies it because he takes pleasure in it; and he takes pleasure in it because it is beautiful."

But this judgment of beauty in mathematics or the natural world seems far more difficult to understand than the processes of reasoning that bring us to such a judgment. Judging a formula, an insight, a method to be beautifully constructed is asserting that the mathematical argument has a quality or purpose quite apart from the reasoning processes that gave it birth. It is as if we were judging it the way the Deity judged His own creations after each day of labor: it is good. We don't need to assert divine intervention in order to experience this wonder. What is important is that at the very moment such beauty becomes apparent to us, the world appears *as if* it were constructed according to some purpose. Or better: the world appears as if it were well suited to human perception and understanding, as if nature were constructed specifically for our contemplation. Kant referred to this as regarding nature "after the analogy of art."

It might seem peculiar to be bringing in such indirect and imprecise sentiments in talking about a subject which insists upon clarity and precision. But these feelings are related to the reasons theories are constructed in the first place. A theory must create order, define conditions, show the relationship between parts and whole. It explains things, but it also discovers things, revealing unexpected connections. "You must have felt this too," the great physicist Werner Heisenberg said to Albert Einstein, "the almost frightening simplicity and wholeness of the relationships which nature suddenly spreads out before us and for which none of us was in the least prepared."

The mathematician J.W.N. Sullivan wrote, "Since the primary object of the scientific theory is to express the harmonies which are found to exist in nature, we see at once that these theories must have an aesthetic value. The measure of the success of a scientific theory is, in fact, a measure of its aesthetic value, since it is a measure of the extent to which it has introduced harmony in what was before chaos." Further: "The measure in which science falls short of art is the measure in which it is incomplete as science."

The mathematician Roger Penrose in *The Emperor's New Mind*, an unusual book on the cosmos and mathematics, even suggests that the aesthetic qualities of an insight are closely bound to its validity. He believes that there is a close connection between beauty and truth. The theoretical physicist Subrahmanyan Chandrasekhar has argued, for example, that the beauty of Einstein's general theory of relativity is partly due to the ways it has turned out to be harmonious with laws and concepts seemingly unrelated to it. So impassioned was the theoretical physicist Paul Dirac with the importance of mathematical beauty for physical theory that he insisted it take priority over truth: "It is more important to have beauty in one's equations than to have them fit [the] experiment." Experiments, after all, could turn out to be mistaken.

DESPITE DIRAC'S FAITH, it is all too easy to think of beautiful things that do not seem true and true things that are not beautiful. Some physicists have even argued that the natural world may be much messier and more inelegant than our theories make it appear. John D. Barrow has written that "there is no real evidence that Nature is in any well-defined sense 'simple' or 'beautiful.' Indeed, the hallmark of most natural phenomena is a deep complexity masquerading as simplicity." And Bertrand Russell once suggested that we may be glimpsing only a part of the universe (that part which can be expressed mathematically) when we present our physical laws in mathematical form.

Beauty is also no guarantor of truth. It is quite possible to imagine a beautifully reasoned proof that is false. There have been exquisitely organized physical theories that became unusable. The Ptolemaic vision of the universe, for example, was based on an elegantly simple assumption: that all heavenly bodies move according to nature's most beautiful form, the circle. The theory faltered and then collapsed as observations forced the addition of intricate circles looping on other circles in attempts to describe celestial meanderings.

Nevertheless, there is something about nature, art, and the analogy we make between them that seems essential to mathematical thinking. For all the philosophical debates over the reality or "epistemological status" of mathematical objects and arguments, most mathematicians regard their enterprise as if it involved treating nature "after the analogy of art." A satisfactory exposition of mathematical ideas is deemed to be one that finds the hidden reality, the essence, the Form. It doesn't even matter what the material is: cups of wine, arrays of numbers, snippets of a line, doughnuts in space, even theories of chaos and catastrophe. The point is to articulate the basic underlying elements of a situation, find analogies with worlds already understood, and open new regions for exploration. Structures are identified, abstractions created, mappings made. Mathematics creates models that reflect the hidden patterns of the world. In turn, mathematics becomes a model of the world it describes; the analogy of art that it establishes for nature can be extended finally to itself. A mathematical model should make the world it describes look as good as the model itself.

This aesthetic imperative has an impact on the substance of mathematics; it does not just affect its surface. Indeed, if mathematical reasoning completely ignores the world which gives it ground and merely spins fantasies upon fantasies, it ceases to be aesthetically appealing; it becomes ornate, unconvincing. The contemporary versions of the Chinese remainder theorem quoted in the second chapter are examples of how mathematics can create filigree; the abstraction has stopped revealing similarities and started to obscure them. By contrast, if math completely ignores the analogy of art, if it is just involved in concrete answers and realistic description, if its intentions are to "solve problems" or provide tools, the results are too particular: they will lack power and generality; they will be useful but hardly compelling.

This imperative also has nothing to do with whether or not mathematics is "applied" or "theoretical"; it has to do with a way of thinking. In the problem with the cup of wine, for example, if we laboriously calculate the proportions of wine and water in a particular case—if we restrict ourselves too loyally to the facts presented—then we may satisfy ourselves of the truth of the solution but we will have no *general* understanding of why the answer makes sense. It is only when we attempt to define the essential aspect of the problem—to create a sort of model for the process we are examining—that we reach the truth with an answer that has elements of beauty.

Kant noticed a quality in aesthetic judgment that we may find relevant in mathematical thinking: the judgment of beauty is removed from any considerations of interest. It is a detached judgment. It has nothing to do with the usefulness of an object, or its monetary value. The beautiful object may inspire a desire to possess it—or understand it—but the judgment of its beauty is untainted by other concerns. Whether we gaze at a sunset on a mountain's peak or listen to a Mozart piano concerto, whether we are enraptured by a person or contemplating Newtonian mechanics or quantum physics, insofar as we make an aesthetic judgment, we have no other interests. Aesthetic judgment is as detached and as devoted to abstraction as is mathematical thinking.

Moreover, the judgment of beauty has an absolute aspect to it; beauty seems well suited to be a companion to truth. When we are gazing at an object, we do not consider our judgment of its beauty to be a mere assertion of taste; we do not emphasize the relative aspects of our experience—the time of day, the particular moment in cultural and historical life, our passing mood. Instead, we feel beauty to be universal, beyond the reach of time or fashion or individual whim. The experience of beauty refuses the particular even as it embraces the abstract; the instant becomes an instance of something else.

There is something about beauty that is both private—because it involves silent feeling—and public—because it makes us feel as if it is revealing something universal. It emphasizes both our isolation and our feelings of common sense and sensibility. For the same reason it can also risk inspiring contemplative withdrawal or impassioned absolutism. The judgment of beauty is not idiosyncratic—or so we think and feel—but something more fundamental. Beauty feels like an aspect of public knowledge. We may not actually assert that everyone *will* agree with our proclamation of beauty, but beauty inspires a feeling that everyone *should* agree. The feeling makes a claim not only on us, as we view the beautiful object, but on our sense of others. (This is the origin of snobbery.)

Of course, we haven't proved that taste is universal or that beauty is objective; the only assertion is that the judgment of beauty is *treated* as universal and *felt* as objective. One of the peculiar aspects of modern aesthetic taste is that the universal feeling beauty inspires is often denigrated as if in fear of its claims; instead, we force ourselves to treat all notions of beauty as being *relative* assertions of taste. The result can be an art that is anxious about its aesthetic quality, as if asserting its independence from all claims, an art that refuses to be public or even beautiful.

Mathematics is precisely the opposite: it stakes claims. It has less to do with feeling than with fact, but in its disinterested contemplation, in its generality, in its attempts to define essential relations and structures, in its individuality and universality, mathematics seems to be an intellectual activity with very close connections to aesthetic judgment. When we contemplate the beautiful we feel that the object has a "purposive" aspect even though it may have no particular purpose, we treat it as if it were specifically constructed for our contemplation, and we sense a resonant or harmonious relation between the object and our unspoken sense of things. We attend to detail but only as part of a whole; we see connections between details, create patterns, then seem to leave all conceptual labor behind for the sake of simple contemplation. Mathematics often inspires such reactions as well. The feeling of beauty is of the universal revealed, similarity disclosed, order constructed: it seems echoed *by* mathematical reasoning and echoed *in* it.

The following example will enable us to characterize one type of beauty more precisely. Take the familiar special symbol used by the Pythagorean school: the five-pointed star. This star, a classic geometric figure, is constructed by the diagonals of a regular pentagon.

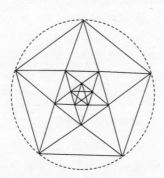

Look at it closely: the diagonals create a second regular pentagon within the first. They also create a large number of congruent (meaning identical in size and shape) triangles. If the second pentagon's diagonals are also drawn, the pattern repeats itself.

There is something enormously attractive about this figure. It has stability and balance both in its overall shape and in the ways the lines divide one another. It also is interesting because of the hints of instability in it, the way alternate stars are precariously perched on points. This star—the pentagram—is thought to have been a sign of the Pythagorean brotherhood; it later became a mystical sign for the human and divine form. The lines of the star, the diagonals of the pentagon, divide one another in a remarkable proportion that Euclid called the extreme and mean ratio: "As the whole line is to the greater segment, so is the greater to the less." This means that there is a relationship between parts of the line that is reproduced in the relationship between the whole and the part. It can be seen in any number of lines in the diagram.

This ratio is unique and can be calculated. If we take a line divided into two parts, a and b, then the extreme and mean ratio of this line would require that

$$\frac{a}{b} = \frac{a + b}{a}$$

We can rewrite this equation:

$$\frac{a}{b} = \frac{\dfrac{a}{b} + 1}{\dfrac{a}{b}}$$

and simplify it by letting $a/b = \Phi$, leaving us an equation

$$\Phi = \frac{\Phi + 1}{\Phi} = 1 + \frac{1}{\Phi}$$

or

$$\Phi^2 - \Phi - 1 = 0.$$

This equation is easily solved; Φ is equal to

$$\frac{\sqrt{5} + 1}{2}$$

This might seem a fairly "unattractive" number given the unusual character of the ratio, which by definition carries itself within itself. What, after all, is so special about $\sqrt{5}$? Besides, this irrational number written in decimal form has all the seeming randomness of π. It is equal to 1.61803398875 . . .

Yet if we multiply this number by itself (squaring it) we get a number that seems related to it:

$$\frac{\sqrt{5} + 3}{2}$$

And if we take its reciprocal ($1/\Phi$), we get another seemingly related number:

$$\frac{\sqrt{5} - 1}{2}$$

In addition, as the original equation told us, if we add 1 to this number, it is equivalent to squaring it; and if we add 1 to its reciprocal, we get the original number. Thus, in a direct and intriguing way, the processes of addition to this number seem closely related to its properties of self-multiplication. Moreover, if we take the preceding equations, writing

$$\Phi = \frac{\Phi + 1}{\Phi} \text{ as } \Phi = 1 + \frac{1}{\Phi}$$

then substitute the right side of the equation for the Φ that appears within the right side, we can create a new sort of equation:

$$\Phi = 1 + \frac{1}{\left(1 + \dfrac{1}{\Phi}\right)}$$

If we keep substituting for Φ on the right side,

$$\Phi = 1 + \frac{1}{1 + \dfrac{1}{1 + \dfrac{1}{\Phi}}}$$

we can actually express Φ as an infinite fraction composed entirely of ones:

$$\Phi = 1 + \cfrac{1}{1 + \cfrac{1}{1 + \cfrac{1}{1 + \cfrac{1}{1 + \ldots}}}}$$

Unlike the other expressions for this number using $\sqrt{5}$, this form of fraction, known as a continued fraction, actually displays something fundamental about this number and its properties: its power to contain itself within itself in the simplest possible fashion. While the unusual properties of π are gnomic to the casual viewer, involving arcane and intricate relations between complex numbers, trigonometric functions, and rates of change, in this case there is something more elementary revealed: the replication of the number within itself. Moreover, an intriguing relationship exists between adding to this number and multiplying it by itself. In fact, add any two consecutive powers of this ratio, and the result is its next power:

$$\Phi^n = \Phi^{n-1} + \Phi^{n-2}$$

So a series formed from powers of Φ, a series created by multiplication, is also an additive series, each number being the sum of the two previous ones. The series is thus both

$$1, \Phi, \Phi^2, \Phi^3, \Phi^4 \ldots$$

and

$$1, \Phi, 1 + \Phi, 1 + 2\Phi, 2 + 3\Phi, 3 + 5\Phi \ldots$$

each number being the sum of the preceding two. This series is also related to a famous recurrent series in mathematics (and, as we will see, in nature) known as the Fibonacci series:

$$1, 1, 2, 3, 5, 8, 13 \dots$$

The Fibonacci series is formed by the same rule. And, it turns out, not incidentally, that the ratio between two adjacent terms of this series approaches our ratio Φ, as the terms become larger.

But Φ is not of interest solely as a numerical ratio. The Greeks understood nearly every numerical relation as a geometric relation. If Φ is the length of a line, Φ^2, for example, is the area of a square. Thus the connection we are noting in Φ between additive properties—lengths being extended—and multiplicative properties—areas being created—brings us back to the realm of the eye. We can begin to see how the abstraction of numbers can be related to a more sensory relation, something available to the eye and to more traditional aesthetic judgment.

If we take a rectangle whose sides are in the ratio Φ—a rectangle, for example, of sides Φ and 1—the nature of the ratio means that if we cut a 1 $\times$ 1 square off the rectangle we will be left with another rectangle whose sides are in the same ratio. The rectangle $ABCD$ has the same proportions as the rectangles $aBCd$, $fECd$, and $fecd$.

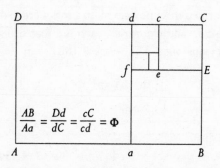

$$\frac{AB}{Aa} = \frac{Dd}{dC} = \frac{cC}{cd} = \Phi$$

Thus the ratio's numerical relations are seen as physical relations, for this cutting can go on infinitely.

This geometric form, like the numeric ratio, has been deemed so beautiful over the centuries that it has been considered "golden." The Greeks made it the foundation for the design of the Parthenon, which reproduces it in its internal proportions as well as in its overall shape:

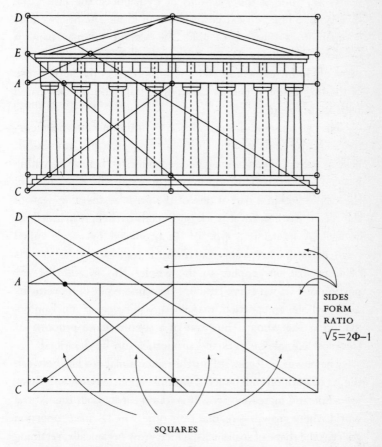

Each small rectangle labeled $\sqrt{5}$ *has sides in that ratio. Note that*
$\sqrt{5} = 2\,\Phi - 1$

When the Renaissance rediscovered Greek architecture and art and the doctrines of proportion became central, the rectangle and ratio became a reference point for painters and architects. Leonardo da Vinci's friend Luca Pacioli called Φ the Divine Proportion. And later Johannes Kepler made it central even in his studies of the heavens. "Geometry has two great treasures," he wrote, "one is the Theorem of Pythagoras; the other, the division of a line into extreme and mean ratio. The first we may compare to a measure of gold; the second we may name a precious jewel." The Divine Proportion has been traced in the plans of Gothic cathedrals and, even, more recently, in the work of the Impressionist painter Georges Seurat, who was fascinated with its properties.

The eye senses in this proportion a continuous internal recurrence. It also finds stability in its dimensions, a piquant restfulness, as the ratios can be imagined reproducing themselves within it again and again. This notion of proportion and replication of the whole within a part is one of the great aesthetic appeals of the recent mathematical work on fractals by Benoit Mandelbrot as well. A fractal is a type of set produced by a rule called recursive—one keeps applying the same transformations to parts of a set that one applies to the whole. This means that any portion of a fractal curve contains the same types of movements as the whole; any portion, magnified, will reveal as much information as the whole. Thus, out of a simple set of proportions the most complex curves and properties can be described.

The power of Φ goes still further since it makes a link between the act of addition and the act of multiplication, between lengthening a line and increasing area. It is found throughout the natural world where growth is regular over time. The Divine Proportion governs the shape of snail shells, which grow organically, retaining a similar shape while increasing in size; it governs the turns of leaves on stems, the arrays of seeds in sunflowers, even the proportions of the human face. It is recurrent in organic forms. (It almost never occurs in the inorganic world, where addition is more important than self-generating development; crystals, for example, are based on different proportions, often on the hexagon rather than the Pythagorean pentagon).

WHAT IS THE RELATIONSHIP between this ratio and the beauty we have been attempting to explore in mathematics? In *The Geometry of Art and Life*, Matila Ghyka, before providing a subtle analysis of the Divine Proportion and other ratios, described the nature of ratio itself. Ratio, he noted, is different from proportion. It means that a comparison is being made. It is "the *quantitative comparison* between two things or aggregates belonging to the same kind or species." In other words, a ratio is an *internal* comparison within a given universe. It involves a mapping between two objects in that universe and a number. We can have a ratio between lines (measuring length) or spaces (measuring area); we can have a ratio between dangers (measuring probabilities of disaster); we can even have a ratio between sounds (measuring frequencies). These worlds of lines and dangers and sounds are mapped into the world of numbers. A ratio, then, requires a "space" of objects, and it requires a measure that we use to comprehend these objects.

Thus, establishing a ratio is already a sophisticated act of the mind. It requires comparison between one object and another, a measurement of their relationship, and an assessment of their differences and similarities. We can go further: it is no accident that creating a ratio is nearly the archetypal act of rationality. It is an act of reason that is not divisive but connective, showing relations between distinct objects.

Because creating a ratio means giving a number to these relations, it also allows us to compare different worlds, not just objects within a single world. We can identify ratios from different worlds through their numbers. The ratio of 2, for example, could apply to lengths of a line, populations in various countries, or the frequency of pitches on a piano. We can say that the ratio of the lengths of two lines is equivalent to the ratio of two frequencies of vibration, thus making a connection between the lengths of vibrating strings and pitch. When compared in this way ratio becomes something more profound; it is being treated as a *proportion*.

A proportion can show the similarities between different worlds; it can demonstrate similarities and relationships that are shared; in a sense, what we have been attempting to find in our explorations of music and mathematics are shared "proportions." The proportion, rhetorically, is an analogy, which has its roots in the word *analogia*, originally used to refer to proportion. As Ghyka pointed out, first the Greeks and Romans and then the Gothic and Renaissance builders used repeated proportions—analogies—in their architecture. A ratio between dimensions was never accidental and always echoed other ratios throughout the plan. Structures contained internal resonances, echoes, and repetitions. The term *symmetry* meant much more than simple mirror identity between two parts; it referred to repeated proportions.

Vitruvius, the great Roman theoretician of architecture, wrote, "Symmetry resides in the correlation by measurement between the various elements of the plan, and between each of these elements and the whole." Vitruvius compared the symmetry of building with the symmetry of the human body: "It proceeds from proportion—the proportion which the Greeks called *analogia*—[it achieves] consonance between every part and the whole." Proportion, he explained, "is a correspondence among the measures of the members of an entire work, and of the whole to a certain part selected as standard." A building built with proportion and symmetry in mind is generated by that standard. Vitruvius defined the aesthetic importance of such proportion: "When every important part of the building is thus conveniently set in proportion by the right correlation between height and width, between width and depth, and when all these parts have also their place in the total symmetry of the building, we obtain eurythmy."

This notion of symmetry and eurythmy is close to what we call, in other contexts, harmony. The Divine Proportion is a powerful example of eurythmy because of the clear relationship it displays between the whole and the parts. It may even have been a Pythagorean belief that the proportion extended from the human body (which, as the Renaissance artists and the ancients knew, tends to be divided into "divine" proportions by the navel) to the universe itself (which was supposed to have a structure like one of the five regular solids, the dodecahedron, itself suffused with divine proportions, like its two-dimensional relation, the pentagon). In later centuries, the analogies between parts and other parts, and between parts and the whole, became the foundation of mystical beliefs upon which the notion of magic depends. In magic, one object becomes an analogy for another (like a voodoo doll for the human body); act on part of one object, and the magic acts similarly on the other. "Metaphors for us, but not for Him" is how, we recall, one of the Jewish mystics explained the manner in which similarities and comparisons in the earthly realm are mirrors of divine equivalencies. Proportion, in these beliefs, can become a magical identification of parts and influence.

There have been many attempts to describe or identify the beauty of different types of eurythmy. Ghyka and other writers have made a distinction between ratios based on irrational numbers like $\sqrt{2}$ or Φ and ratios of rational numbers like 2 or 3. The rational ratios, while ostensibly simpler and more elegant, offer fewer possibilities for internal echoes and analogies. The irrational ratios offer greater complexity on the surface but more profound similarities below. Even the Greeks understood that those irrational relations created more dynamic shapes, with greater interaction between parts and whole. Because the irrational proportions are not easily divisible into simple compartments, because their ratios imply an infinite sequence of relations, the eye and the mind are pulled to their geometric representation (which suggests that beauty may sometimes require a more profound simplicity than one presented on the surface).

It is possible to analyze a form to determine what "types" of proportions and ratios are displayed within it. One scholar established a typology of architectural forms in ancient and Renaissance architecture, an attempt to articulate basic "themes" of relationship and proportion that underlie much of Western art. It resembles the attempt by Heinrich Schenker to find the fundamental melodic line underlying Western tonal music, those archetypal descents ③-②-①. These themes are based on properties of geometric ratios—ways in which rectangles or other shapes can be divided, ways in which parts relate to wholes.

Ghyka calls these, and other diagrams of relations, "harmonic analyses" that show the "thematic" consistency of such "symmetric" and "eurythmic" figures. Here, for example, are a few such "harmonic analyses" for rectangles with sides of length 1 and $\sqrt{3}$, and 1 and Φ; ratios of $\sqrt{3}$ and Φ reappear inside the rectangles' divisions.

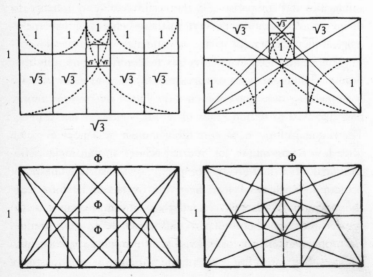

These notions resonate with our experience of mathematics. In mathematics we create patterns that make sense out of what we are presented; we create constellations that are abstract organizations of details. Each constellation represents a multitude of cases within it; and the constellations are, in turn, gathered into still more abstract entities. We speak of beauty in mathematical argument when those constellations show us relations we have never seen before, when we have found in them repeating and resonating essences and forms. In doing mathematics we are, at least metaphorically, finding proportions, identifying ratios, creating mappings between different instances of the same fundamental rational pattern, linking different worlds by finding their similarities.

Gauss, in this way, essentially found a harmonic analysis of arithmetic series. Euler found a kind of harmonic analysis linking π and e and i, as if the essences of these numbers and their relations were being revealed, a hidden proportion found. Mathematics is always using analogy and proportion in its argument. In fact, when faced with a form—whether it is musical or mathematical or even in the physical world—we attempt to make analogies among the form's parts. We find in these analogies and internal mappings the fundamental proportions governing the form. This is how we memorize telephone numbers, looking for patterns even in randomness; it is how theories are constructed; and, as we have seen, it is how we come to understand music as well.

Ghyka called this process the principle of analogy; it is, he noted, "common to Art and Science. *Analogia* itself, the geometric proportion (A is to B as C is to D), was the capital tool of Euclidean Mathematics." Analogy, he explained, is at the base of eurythmy and modulation, in music and in physics. It is also part of the art of literature, which depends for its power on metaphor, which is a "condensed and unexpected analogy." He quoted Aristotle: "The greatest thing of all is to be a master of the metaphor. It is the only thing which cannot be taught by others; and it is also a sign of original genius, because a good metaphor implies the intuitive perception of similarity in dissimilar things."

This quest for the appropriate metaphor, for proportions and analogies governing the structure of nature and art, for mappings which reach the essential elements of the objects we study—this quest deals not only with objects but with processes. In coming to understand something or make sense of it, we must be attentive to "proportion" and analogy in their most general sense. We must create metaphors as we work, exercising, in Aristotle's phrase, the "intuitive perception of similarity in dissimilar things." This is, in essence, the activity of mathematics.

That activity also has something to do with our experience of beauty itself. It touches on the mechanism by which we come to know beauty and "feel" the aesthetic. What we "feel" at such moments is the analogy of part and whole, object and other object, relation and relation. This is one reason that in moments of aesthetic transport we assert the universality of the beautiful: we are feeling something not inchoate but precise and seemingly beyond contradiction. This is also why we feel something similar when coming upon the beautiful in art (or music) and the beautiful in nature (or mathematics). As Schopenhauer asserted, "Aesthetic pleasure is essentially one and the same, no matter whether it is evoked by a work of art or immediately produced by the contemplation of nature and life." Beauty is experienced as a form of knowledge because it is through the archetypal rational act—that of analogy and metaphor—that we come to know the beautiful.

WE HARDLY NEED to labor as fiercely to see similar patterns in our comprehension of music, for much of what I have said already applies to music without major elaboration. Beauty is not an esoteric property of musical composition; the ambition to achieve it is often the very point of music. But we cannot simply carry over our earlier considerations about beauty, for in any mapping between systems, any analogy we set up, there are the incidental similarities and the essential ones, the differences that have no overriding significance and those that define alternative worlds. So it is worth stepping back for a moment the better to comprehend the nature of musical beauty.

As we have seen, when we observe the natural world and are moved by it, whether by studying a set of formulas on a paper, tracks of particles in a cloud chamber, theoretical intimations of black holes and singularities in space, or dawn at the peak of Mt. Snowdon—when a scene unfolds for us elements we did not expect, ripe with relations, revealing objects and their connections—we think that here is nature displaying its purpose. We see not its literal purpose but its *purposiveness*. Depending upon our religious predilections, we may take this either as a literal assertion or as a conceptual analogy created by the observing mind.

When we confront a work of art, of course, that purposiveness is not at all *seeming*, it generally *is*. The work has been made to create a particular effect, and the creator can hardly be ignored. The purpose can vary widely and may have nothing to do with artistic self-display or self-expression (though it is easy in contemporary culture to forget this point). There are cases, for example, when the artist deliberately hides himself (as in the great Gothic cathedrals): the object then is to present the artwork as if it were being offered on the model of nature itself, beyond human ability, participating in some order that lies beyond the individual. But in many cases, the observer is meant not to marvel at the inventiveness of the artist but to react to the work as if it were hardly constructed by human agency at all. The religious powers of art rest on such a sentiment; the artwork becomes transparent and speaks as nature.

We can have such an immediate reaction to the aesthetic creation lying before us, such awe at its internal workings and the powers they exercise upon us, that the work seems something not made but simply presented; it requires no analysis of purpose nor any reflection about meaning; it appears to our senses as something completely natural. If in mathematics we treat nature as an analogy of art, in the greatest of music we treat art as an analogy of nature.

Even in works suffused with the ego of the creator—such as the symphonies of Beethoven—this sense of nature and natural force can be profound. The work can seem both inevitable and surprising, both totally natural and extraordinarily unique. We can lose our selves in such music because in the act of reflection we are engulfed, not as mere observers but as participants in another world. "The imagination," Kant observed, "is a powerful agent for creating, as it were, a second nature out of the material supplied to it by actual nature." In music, we can dwell in a second nature.

Apart from the act of composition, it is in the playing of music that its powers become most clear. At any moment there are countless activities occurring: notes being read, fingers or breath to be controlled, harmonies sensed, voices articulated. Anybody who has ever attempted seriously to master an instrument must remain humbled by the recalcitrance of body, mind, and ear in the face of difficulties that expand the more one progresses. Yet after the body has become thoroughly accustomed to the extraordinary artifice of the exercise demanded of it, the physical difficulties dissolve. There can be a feeling of intoxication, as if something is being spoken "through" one, and all one need do to uncover still more of that universe being revealed is pay attention.

At this point in our experience of the artwork we have already moved beyond the knowledge of analysis, or the information gained by reflection. We are not talking of the formal relations that exist within music; we are not talking about the way music works. We are talking about the way it makes us *experience the world*. The testimony is irrefutable, if imprecise: music has the power to change the way we see things, to transform our senses and our understanding: it presents itself not as *a* creation but as *the* creation. This feeling is inextricably linked with the sense of beauty we have been exploring; somehow the internal relations of the composition and its external relation to our experience become related. We are in another world, but through that world we seem to find our own.

How does a work of music inspire such feelings? The phenomenon is more obvious here than in mathematics, perhaps, but its essentials seem harder to understand. We cannot just point to the use of repeated ratios or the Divine Proportion in music. It has been argued, for example, that Béla Bartók was so haunted by that proportion that he would plan his compositions around it, reaching a climax at the precise measure which divided a composition's duration into this familiar ratio. But while this fact is intriguing, and sheds some light on Bartók's ambitions, it does little or nothing for ours. We don't experience ratios of time as we do ratios in space. If we have a composition in which one section is half as long as the section following it, we don't experience a ratio of 1:2. Time is not accounted for the way space is but as a uniform measure.

What is essential in music is the *experience* of time we have when listening, and this is a complex matter indeed. A measure of sixteenth notes seems to move much faster than a measure of half notes. And the experience of something later is necessarily affected by what transpired first. What is the ratio between the musical durations, for example, of a sonata of Mozart and a mass by Byrd? This is not a simple matter: musical duration is not found through calculation, nor can musical durations—that is, the temporal experiences of compositions—be simply compared; the experience of time is too different in each work.

It is almost as if different measures were being invoked or no real measures at all—like asking for the ratio between a raven and a writing desk. Our analogies must be more sophisticated.

If we are to draw any closer to musical beauty, we must take music on its own terms and respond to its *internal* life. We must recognize that our sense of beauty in music involves an element seemingly absent in mathematics: a sensuous presence. There can be something beautiful in the sound itself, even if there is no musical order surrounding it. I have known pianos, for example, which have had such beautiful timbre that it was possible to play a note and listen to it slowly decay into silence, relishing its pointed beginning, its warm early resonance, and the sighing of its decrescendo as overtones slowly dissolved. A harmony can also be beautiful: a chord can be beautiful as a sound, not just as an element in a composition; it too has a sensuous presence. And we need not take the soft-edged or subtle as representative of beautiful sound; there is something about the biting, melancholic vigor of the harmonies in Bartók's string quartets that is sensuously exciting as well; their defiance and edginess are often grating yet also quite beautiful.

Yet what is it that makes a sound beautiful? Would any chord really sound beautiful if it did not have the network of implications and tensions analysts have spent decades defining? Does an instrument sound beautiful in itself without that sound suggesting a sort of musical potential, an implicit possibility of elements that can unfold from the vibrations being heard? The great performers I have heard play—such as Artur Rubinstein and Herbert von Karajan in their very last years—created a sound that was inextricably linked to what was made with that sound: the creation of the sound was a musical project, not a purely technical one. The sound was resonant in the sense that it had implication, complication.

It is also possible, of course, for beauty to be terrible—a

possibility to which we shall return. But the beauty of musical sound—even the beauty of sounds that seem purely sensuous, physical events—also involves the notions of proportion and analogy. When Rubinstein played the "Funeral March" from Chopin's Bb-minor Sonata during one of his last concerts, the sound was chilling, but its beauty lay partly in the way the sound's pianissimo haze seemed almost to create a tactile substance in the air, a cloud hovering over the phrases. The sound created analogies in the ear: Rubinstein used it structurally to subtly alter the piece's texture as the march grew more insistent. It was also, despite its delicacy, a rich sound; it had dimension; one could listen *into* it. It seemed to have shape.

The beauty of a sound is like the beauty of a shape: it must seem to have an internal harmony, which is a subtle balance between high overtones. It is in the relationships of those overtones that we hear the greatness of a violin or a piano. If a sound is stripped of some overtones—as many electronically produced sounds are—it can still seem to have beauty, but a beauty that is usually described as cold. If a sound doesn't blossom with implication, it is stillborn. For the same reasons, the sounds which are generally created by compact disc players are not beautiful: they are just technically polished. Compact disc sound usually lacks the nuances and subtle body that can be found in the sound of real instruments and were more fully captured in fine long-playing records. There is no presence in this electronically manufactured sound, only the sound of absence.

Hermann Helmholtz ends his classic study on the physics of musical sound with a section that discusses the relations between complicated sounds. He writes:

We recognize the resemblance between the faces of two near relations, without being at all able to say in what the resemblance consists, especially when age and sex are different, and the coarser outlines of the features consequently present striking differences. And yet notwithstanding these differences—notwithstanding that we are unable to fix upon a single point in the two countenances which is absolutely alike—the resemblance is often so extraordinarily striking and convincing, that we have not a moment's doubt about it. Precisely the same thing occurs in recognizing the relationship between two compound tones.

The relation between two tones has a sort of uncanny quality in their combination as well. Helmholtz even compares the sense we have of relation between different intervals with the sense one has in looking at an artist's creation of resemblance in a portrait, suggesting that the artist must possess "an unusually delicate feeling for the significance of the human countenance." The suggestion is of sounds that resemble each other in inarticulate and barely conscious ways, as if they were simply providing an occasion for the mind's own associations and innovations. This is partly why one feels one's mind so active when confronting music and harmony. When we hear simple intervals bearing such relations—the octave and the fundamental tone, for example, or the third and sixth—we hear something that is both alike and unlike. These intervals are beautiful in the way the shape of the circle can be considered beautiful in itself, or the way the conic sections—the parabola, ellipse, and circle formed from slices of a cone—can seem hauntingly related.

The beauty of individual sounds and their combination has something to do with music's origins: it is out of timbre and texture that melodic beauty grows, responding to latent proportions and tensions and dissonances. The Bach fugue mentioned in the first chapter is so affecting partly because its theme grows out of the interval of a fifth—the most consonant of intervals smaller than an octave, with the most overlapping overtones.

But the movement of the line immediately turns against that stable, open sound; it already has sorrow in its countenance by the third bar. When the melody reaches up yet again, with the interval of a fourth, we experience this second reach as a response that recalls the one just past and continues it. This second interval is given sense by what precedes it, and by the way it both replicates and varies what came before. It is less vigorous, more resigned, a sign of introspective descent. Those leaps of the fifth and the fourth, in turn, determine the entrance of the other voices of the fugue.

Another beautiful melody is Cherubino's love song to the Countess in Mozart's *Figaro*: each movement has a countermovement, every downward sigh has an upward strain. Its simple texture and repeated gestures, its leaps and collapses suggest a mixture of innocence and desire, engaged in a delicate dance— a model of Classical symmetry in sound. I could make this description still more precise by pointing to ways in which the line flirts with delayed resolution yet moves forthrightly, with unassuming simplicity.

When we think of beauty in music we have to think of such relations and references, the sort of network of implications and recollections we saw at work quite formally in the last chapter. A composition is a construction of patterns and proportions, resembling an argument in mathematics. Like the mind that makes sense of mathematical progressions of argument through examining the invocations of principles and assumptions at each step, the ear maneuvers its way through a composition. The ear follows time but is also constantly moving backward as it moves forward. For the only way we come to know music is through comparison, through the noting of differences, similarities, and transformations.

The judgment of beauty in music has something to do with this experience of internal relations, the experience of connection and variation. Variation can mean simple change, but it is also the fundamental principle by which music is created. It is, as we have seen, the musical version of mathematical mapping. The variation is only effective, though, if there is some essential or basic kernel, felt in our ears and our minds, something invariant, an unchanging substratum that allows us to recognize what is being changed. A simple example: near the opening of Beethoven's *Pastoral* Symphony, a single phrase is repeated over and over with the same accompaniment; the only change is the instrument playing the phrase and its volume. This variation gives us the experience of transformed timbre and increasing intensity. It focuses our attention. It also gives us a sense of essence lying under the variation. But, paradoxically, even an unvaried repetition in music is a kind of variation, for the passage is not heard a second time in precisely the same way it is heard the first.

The musical variation involves a process of abstraction: it examines a phrase of music and splits it into its essential elements: a motive here, a rhythm there, a harmony, a gesture. These elements are either held constant or varied. A variation usually concentrates on a single transformation. The musical theme and variations, then, is not just one musical form: it is *the* musical form. It is, rather, that which allows music to be formed. Whenever we are asked, in music, to feel some connection between one moment and the next, we are asked to attend to elements that remain the same and those that change. Variation is music's mode of argument. It establishes similarities and analogies. We know two moments are analogous when there is enough similarity between them to create a sense of recognition. These moments of analogy connect not just such instants but also the regions surrounding them. Music is structurally ordered by analogy.

A good example are the Mozart variations to which we referred earlier, based on the tune we know as "Twinkle, Twinkle, Little Star." Most of these variations keep the theme clearly present and vary the rhythmic aspects of the accompaniment.

Mozart's variation on "how I wonder"

As the set of variations reaches its climax, there is a slow variation which is nearly a fantasy. The plodding descending line of "how I wonder . . ." becomes typically Mozartean: it takes a step down, then leaps up only to sigh its way down and begin again, a bit closer to its goal. The essential elements of the melody are still present; they give the variation its shape and structure. But the variation leaves us with a sense of something latent in the original that we may have missed—the sighing descent of the line, with its halting movement.

Such a simple variation provides us with a clearer view of the object being transformed. We may find it particularly beautiful when the discovery we are granted about the musical material has some of the elegance and compactness we saw in mathematical argument. Beethoven is often startling in this respect: frequently, in his work, a musical phrase will suddenly gain new significance as it is seemingly derived out of another. Even individual notes can be reinterpreted. A jarring, disturbing C♯ in the opening theme of the *Eroica* Symphony reappears during the recapitulation, in which the C♯ is replaced by the equivalent D♭, which leads downward rather than upward. At moments like these, we come to know the original object, the theme, better, but we also come to know the mapping, the transformation that shows it in a new light. In Mozart's variations, the step-by-step descent of "how I wonder"—almost a Baroque gesture—is elongated, each sigh ornamented to reveal an intricate inner life on the verge of Romanticism. We experience the relationship between those two phrases, and we become aware of a process that has changed one into the other. Such variation brings us closer to the elemental nature of style itself, which determines what can be varied and what cannot, providing the premises and the kernels of musical thought, the rules for transformations.

Beauty in such a compositional universe can have some of the same origins as beauty in mathematics. Following the premises established by a composition, we can suddenly be brought to an unexpected musical place; we can experience a large-scale transformation that echoes the smallest detail. We can even experience something like proportion.

In mathematics we know we can assert that a torus is topologically equivalent to a coffee cup; we are defining which aspects of each surface we find essential and which aspects are irrelevant; we are reducing each shape to its formal purity even as we define the elaborate surface differences. We are actually establishing a transformation of one to the other; the coffee cup is a variation on the torus, a mathematical variation. To put it another way, when we establish such variations, we are defining a type of ratio between the two objects, a comparison in mathematical space.

In music we do something similar when we compare two musical phrases and experience a transformation between them. This is not just a metaphor. The two phrases might be identical but for their key: one statement may be in C major, the other in F. If we consider a "ratio" between the two phrases, we look for the defining aspects of their differences and similarities. In this case, the ratio between the second and the first might well be considered the only difference, the relationship between the subdominant key and the tonic key: IV to I. In fact, the very analysis of music depends on such ideas of ratio. Naming a chord a I chord, or a V_7, is not saying what the chord is specifically— we have no idea if it is founded on an A or an F♯ or what particular notes are in it; the name does, however, define its harmonic ratio to other chords in a composition's tonal space.

A ratio between two keys might also reappear at other moments in the composition, setting up other sorts of parallels. When we find a movement of keys from C to F (I to IV) at one point in a composition and from E to A (III to VI) at another, we may hear that second example as a movement from I to IV as well, just starting at a different place. (Analysts might even write the new harmonies as I/III and IV/III, since E, the third degree of the scale beginning on C, has become a new tonic, the basis for all harmonic comparison.) The transitions may differ in details, but they serve similar functions; there is an analogy between them; they establish equivalences and connections. Other sorts of ratios might establish analogies between duration, rhythm, or emotional tenor. We also might ask more general questions: Is section A related to section A′ in a piece the same way B is related to B′? Is A related to the whole the way the opening measure is related to the first phrase? Is A's contrast with Z similar in any way to D's contrast with X?

Proportions set up structures within a composition; they are the patterns our ears hear in much the same way that Ghyka's rectangle diagrams show internal structures of repeated ratios. They also give each composition its distinctive topology, its sense of internal space; they tell us where that space bends, how far it is from one point to another, whether or not that space is continuous or shattered into distinct and separable regions. Proportions establish meanings.

We cannot fully understand the *Appassionata*'s restless energy, for instance, without hearing those recurrent relations between the C's and the D♭'s, in both the motives and the harmonies. Proportions are established between the small detail and the large-scale modulations, between the structure of the theme and the intention of the work itself. One of the reasons that the sonata form has been so powerful an aesthetic influence in Western music is that it creates archetypal analogies, a set of proportions guiding harmonic and melodic transformation.

When we judge the beauty of a composition, we are partly judging the network of implications and relations the composition establishes through its topology. Do we feel that the moments of analogy are justified? Is there a set of internal relations within a composition that seem to have coherence? Do we feel that a particular event has some connection to another or does it seem imposed according to some alien set of assumptions? Logical judgments are a part of aesthetic judgment.

THERE IS ALSO another, elusive factor to be considered in judging musical beauty. Mathematics may have provided our guiding metaphor, but beauty, for all its supposed timelessness, for all its appeal to universal applicability, for all its sense of transcending material interests, also seems subject to the vagaries of history and time. Indeed, the characterization of beauty that has emerged out of our considerations of proportion might legitimately be thought of as "classical" both because of its origins in Greek and Roman philosophy and because it has become enshrined in the music of the Classical period in the West; in that repertoire, the tonal musical system allows balance, proportion, and the connection between detail and whole to be thoroughly worked out. Nothing is extraneous; even the unexpected is sensible and comprehensible.

But during the last half of the twentieth century, years in which many listeners have found little of beauty in musical composition, and in which some composers, particularly in the avant-garde, have declared little interest in the cultivation of the beautiful, what claim can we possibly make for transcendent aesthetic principles, let alone a standard based on Classicism? We are more inclined today to reject any notion that aesthetic taste has objective reality, let alone that there are universal criteria for judging beauty.

But this conclusion may miss the ways classical beauty works: it does not prescribe or proscribe. It demands instead a way of thinking and feeling, an attitude toward the universe found within an artwork. When these demands are violated or ignored, they still seem to have power; they often seem like Schenker's fundamental line, present even when submerged behind distractions and opposition. We even find them in works from non-Western cultures—though that is the subject of another journey.

Consider Chopin's Prelude in A Minor, which is an extraordinarily peculiar piece, just twenty-three bars long, and beautiful in a disturbing and disturbed fashion: It is quite unlike anything Haydn or Beethoven might have composed. Not only would we be hard pressed to call some of its harmonies beautiful but its very shape seems to belie the arguments about proportion I have been making. André Gide referred to its inexhaustible emotion and a "kind of almost physical terror" it inspired. Lawrence Kramer, in a suggestive analysis, said that "much of it is deliberately ugly by early-nineteenth-century standards, and arguably by ours." In fact, this little composition has received an extraordinary amount of critical attention over the last thirty years precisely because of its peculiarities.

Chopin's Prelude in A Minor

Chopin's piece has a haunting quality, as if it were teetering on the edge of either beauty or dissolution; a silence, or a syncopation, or a sudden familiar harmony gives rise to an eerie world of fluid and unpredictable musings. Its most sensitive commentators have treated the piece as an icon, marking a change in musical sensibility and aesthetic taste. For Leonard B. Meyer the prelude is a lesson in how melodic continuity is created in the midst of discontinuity; for Kramer the piece's "gradually unfolding antagonism between melody and harmony" represents an antagonism between elements of Classicism and Romanticism; and for Rose Rosengard Subotnik, the composition's avoidance of a strong harmonic ground shows how Classical beliefs in universalism and internal coherence are subverted by Romanticism's preoccupation with the "immediacy of the moment."

Even if we try straightforwardly to name the various harmonies in this piece—the way it seems to move from E minor to G and D and then to A minor, we are dealing with not an orderly movement but a treacherous and often ambiguous path. At the center of the piece, its crisis (measures 11–13), it is impossible even to name the chords. Yet naming chords, as we have seen, establishes context and ratio; here ratio has turned irrational. Proportions cannot be established.

We experience that disturbance, of course, because proportion is still the guiding principle. One critic, Michael Rogers, has suggested that the harmonies of the work divide the prelude according to the golden mean. Perhaps. For our purposes, all we need do is understand how we experience this piece when we listen, what ratios we establish, how we make sense of it as a whole and in its parts.

We begin by hearing a plodding, funereal pattern in the bass, which takes on a function quite different from an ordinary accompaniment. Though the bass outlines harmonies and responds to melodic events above it, it clearly has an autonomous life. The bass establishes the breath of the piece, its pace, its periodic rising and falling. It also gives us a sense that we are entering a world that is already in progress. That sense is emphasized by the melodic line that enters in measure 3; it seems to float without clear direction. An E is sustained, then two major structural moves take place: first, its main motion is a drop to D; then, after some ornamentation, the line drops again to a B. This same gesture of descent takes place three more times, beginning again at measure 8, measure 14, and, in part, measure 20. In each case we are hearing an analogy with what has come before. But it is a series of analogies that cause us dismay.

The reason appears when we examine what is parallel in these repetitions. The second descent, for example, begins in midmeasure with a note held less than half the length of the E that we heard the first time; we expect it to be held longer. The repetition is also disturbed by what happens in the seemingly autonomous bass. While the initial statement had some harmonic coherence, this second statement involves unnameable chords—chords that seem to have no clear function or relation to the harmonies surrounding them.

The third time we hear the descending theme it disrupts even the expectations the first two statements have engendered. It enters after an extraordinary delay of one and three quarters measures. The sense of internal pulse—established by the plodding bass line—is undone by the sense of external pulse, which is consistently violated between phrases and often within the treble phrases themselves. During this third repetition another disruption also takes place: we seem to be moving toward a sort of home—the implied key of the piece, A minor—but we have been so turned about by the preceding harmonies that the pending stability is heard as disorienting. The F natural, which should settle accounts, instead jars against the tonic chord of the bass line. And the descending second that has defined the melody has been expanded to a third, making that F natural sound still more peculiar.

The fourth descent occurs at measure 20 and takes the disruptions even further. It enters after a half measure of complete silence—a silence that displaces a harmony we have been led to expect. We are left without a frame of reference; the pulse is gone, and the implications of the repetitions and the harmonies have been ignored. This final statement then enters, but seems to be no more than a fragment, a repetition of the final half of the descending line. But it may also be restoring a kind of order: it begins on the same note that ended the last phrase, just as the second phrase began on the final note of the first (albeit an octave higher). And it unexpectedly completes the composition, accompanied by stable chordal harmonies. There is something uncanny about it. It is also actually a reversal of the first statement; its structure begins with a descent of a minor third and ends with a descent of a second.

In each case there is enough similarity in these descending phrases to make the parallels unmistakable, yet there is also something that undercuts the analogy being made, violates the proportions the music goes out of its way to establish. Kramer has noted that the melody pays homage to Classicism and the harmony to Romanticism; Subotnik has pointed out that we have left behind the universal aspirations of Classicism for a contingent, individualistic world in which there is no accepted norm. Both are correct.

The impact of this piece is all the greater because it is so technically well behaved in its oddities. Each detail, including the pauses and interjections, serves a purpose. At the work's center, for example, there is an often neglected alto voice in the funereal left hand (measures 8–16), whose structure also follows the piece's fundamental melodic pattern, descending by a second and then a minor third, from a D to a C and then to an A—an analogy that gives the fourth melodic descent, from D to A, with its inverted order of minor third and second, still more poignancy. The piece is full of similar details, analogies that force the ear to make connections; they work subliminally, not overtly. Those connections imply an almost Classical ideal, strained by troubling sensations and disruptions; the order allows us to sense the disorder.

While this prelude could hardly be described as unambiguously beautiful, its impact depends on the properties of beauty we have seen in mathematics: compactness, unexpected results created out of seemingly simple phenomena, analogy and proportion used to create greater understanding, a sense of internal echoes and external resonance. But there is something else in this prelude that we have not yet really come across in our explorations of beauty. It is something almost opposed to beauty.

To review Kant: beauty makes it appear as if the object we are gazing at were perfectly adapted to our understanding, as if there were some unexpected harmony between the mind of the observer and the object being observed. Beauty seems to have its home out there, in the world; it has a public meaning. And it seems to make a claim of objectivity; it inspires one to think every viewer should consider the object beautiful. A beautiful object, whether in nature or in art, also seems to have a purposiveness, a sense that it was made specifically for our contemplation. Beauty calms the conflicts of the rational mind; it reveals, enlightens, affirms, charms, delights.

But this Chopin prelude, along with so much other music, often seems to do the opposite. It does not seem to construct an objective world beyond all personal or individual interest. When we listen to it, we are not tempted to assert that each person should react in a similar manner. It seems instead an almost private meditation. And while it is full of internal allusions and relies on structural comparisons, its proportions offer no great satisfaction in themselves; they are almost used to undercut their implications. The predominant impression borders on the inchoate. The music does not delight; it disturbs. The prelude seems to be subjective, internal, tentative. It does not invite us into an objective, external, forthright coherence. To a certain extent, as all its commentators have made clear, it *overwhelms* our attempts to make complete sense of it; it defies us; it confounds us even as it nourishes us.

Subotnik has noted many of these characteristics and found them to be signs of shifts away from the Classical style and the beginnings of Romanticism. True, but these qualities, in varying degrees, are also part of a wider experience of music itself, found in many eras and cultures. The Romantics, for the first time, took these aspects seriously, elevating them in importance and even using a word to describe them. That word is *sublime*.

Sublime is usually used to describe a natural scene that in its immensity and grandeur seems to dwarf the viewer; the sublime, as Kant noted, can be found in "shapeless mountain masses towering one above the other in wild disorder with their pyramids of ice" or in the "dark tempestuous ocean." The sublime is tremendous, awful, and humbling, yet also elevating. In Kant's description, the sublime, unlike beauty, can be found even in formless objects. The sublime is linked to limitlessness and the infinite, yet it also has its effect because that limitlessness is somehow grasped and experienced at once, as a single whole. And while beauty makes it seem as if an object were adapted to our rational mind, the sublime seems to subvert our judgment, perpetrating, in Kant's words, an "outrage on the imagination." It makes the imagination seem inadequate while giving our understanding an almost ecstatic sense of having apprehended what should be beyond its containing powers. The effect of the sublime is not out there, in the world of objects, but in the experience of the subject. The sublime is part of inner, not outer, life.

We can find something like this in mathematics as well;

indeed, Kant identified a type of the sublime as the "mathematically sublime" because it established its sublimity on the basis of immensity. We can find literal examples of the mathematical sublime, as when Cantor defined the concept of different orders of infinity and proved that there is actually an arithmetic governing them. When we contemplate such concepts, we find a perfect example of what Kant described as the effect of the sublime. On the one hand, the mind is humbled by finding its imagination unequal to its ideas; the mind cannot grasp what it can think; on the other hand, it feels itself "elevated in its own estimate of itself" because of the power of such ideas. We can hardly imagine so many infinities, though mathematicians have defined their character and created extraordinary systems using them.

Chopin's prelude is surely not sublime in all these respects; it is brief in duration; its dynamics are not infinitely large, nor are its ambitions. But its effect is not through size but through structure. It defies our imagination to give it coherence even as we grasp it as a whole. There is something in the interior world it creates that seems without boundaries, an "outrage," a subversion of our imagination that is curiously elevating because of the way it inspires a very high estimate of our own sensibility as we listen (something that was Chopin's defining genius). It is an attempt to create a condition in the listener that might be called the psychological sublime, in which the inner life seems to take on the character of a mysterious landscape.

When we turn from this prelude to pieces like Berlioz's *Symphonie Fantastique*, or Wagner's *Ring*, or Mahler's Eighth Symphony, which can accomplish the same feats while overwhelming the imagination with their size, volume, expanse, and seeming formlessness, we find the sublime in full force. "For the beautiful in nature," Kant wrote, "we must seek a ground external to ourselves, but for the sublime one merely in ourselves." These various compositions create art after the analogy of nature, a nature that affects us like the revelation of some alpine scene with rushing waters and gleaming light and shadowed mountain peaks: in it we see reflections of states of mind, images of consciousness.

The search for the sublime links music and mathematics. Both arts seek something which combined with the beautiful provokes both contemplation and restlessness, awe and comprehension, certainty and doubt. The sublime in mathematics and music sets the mind in motion, causes it to reflect upon itself. We become aware first, in humility, of the immensity of the tasks of understanding before us and the inabilities of human imagination to encompass them. The sublime inspires an almost infinite desire, a yearning for completion which is always beyond our reach. But we are then comforted by the achievements of reason in having brought us so close to comprehending a mystery fated to remain unsolved.

V

FUGUE:
THE MAKING OF TRUTH

When judging a physical theory, I ask myself whether I would have made the Universe in that way had I been God.

ALBERT EINSTEIN

THOUGH THE END OF OUR JOURNEY IS IN SIGHT, THE DIFFERENT paths we have taken are still shrouded in mist. Though the earth may occasionally be bright at our feet, we have yet to see sunrise from the top of our Mt. Snowdon, let alone the vision that awaited the poet at the end of his climb. For while we have begun to understand the nature of musical and mathematical thinking and have touched, tentatively, on the nature of musical and mathematical beauty, the most vexing questions remain. Despite occasional glimpses of other regions, we have been preoccupied with purely formal relations, the workings of math and music at the most abstract levels. We know that these constructions involve transformations, mappings, play with resemblance and difference; we know too that in these constructions there are exquisitely unexpected alliances and connections, resonances and proportions.

But these are incomplete speculations. Games of chess can also be beautiful constructions, full of internal allusions and wit, but chess is limited in a way that music and mathematics are not. Chess is merely abstract; the game does not transform the world; a brilliant strategy does not have much impact outside the universe defined by the game. But there is something more than abstract about mathematics and music.

We already know this to be so from our speculations about how the beautiful and the sublime relate to our experience, and how some perception of beauty is related to our understanding of the truth. But our understanding is still imprecise and may be doomed to remain so. Why did the Pythagoreans believe music and mathematics to have connections both inward and outward—affecting the currents of our souls and the structure of the universe itself? Why are mathematics and music often associated with religious ritual—giving it shape and order while attempting to invoke forces seemingly beyond reason? What gives mathematics its continued power in explaining the world in unexpected ways? And what gives music its power to inspire fear and dread and ecstasy in its listeners, leaving not even the most powerful institutions untouched by its influence? These are not questions about the internal workings of mathematics and music but about the ways they map *into* the world; these are questions about their meaning and truth.

THE PROBLEM in mathematics can be posed differently. We know that mathematics gives us tools for organizing phenomena within the world. The puzzle is why mathematics at its most abstract turns out to have such *power* as well. As we have seen, systems of non-Euclidean geometry were abstract mathematical constructions before it was discovered that they could also describe the physical world. For instance, knot theory—a totally abstract exploration of knots—turned out to be an effective tool in a recent theory of the cosmos, known as string theory. Why should this be? What is the connection between brain-spun speculation and brute physical fact?

One physicist, Eugene Wigner, in an essay titled "The Unreasonable Effectiveness of Mathematics," wrote, "The enormous usefulness of mathematics in the natural sciences is something bordering on the mysterious. There is no rational explanation for it." Albert Einstein asked, "How can it be that mathematics, being after all a product of human thought independent of experience, is so admirably adapted to the objects of reality? Can human reason without experience discover, by pure thinking, properties of real things?"

It is difficult to answer such questions when it is unclear even what "pure thinking" might be. There are mathematical truths so powerful and pristine they bear little resemblance to everyday truth; there are other mathematical truths that seem manifestly inadequate for any use. A syllogism may be true— "If Socrates is a man and all men are mortal, then Socrates is mortal"—but it can also be useless; it is just as true, for example, to argue: "If Socrates is a bird and all birds are immortal, then Socrates is immortal." But this is a true syllogism in which every term is false.

We cannot even take simple mathematical argument for granted in its relation to our world. Ludwig Wittgenstein—in later life a radical skeptic about logic and mathematics—argued again and again that in moving from the world of our experience into the world of mathematical reasoning and back again we are making very great leaps. We might feel compelled to come to a conclusion in mathematical and logical reasoning yet feel absolutely no such compulsion in our understanding of the world. Mathematics has a relation to experience that can be as flexible and problematic as that of language itself. It can also take on a status which no other aspect of our lives shares. How, for example, should we treat an ordinary arithmetical statement like $2 + 2 = 4$? Is it the result of an abstraction from experience, as John Stuart Mill would have asserted? If it is a hypothesis deriving from experience with real objects, then it should be subject to verification and proof from experience. But, in fact, if we saw an instance in which 2 + 2 were not 4, we would discount our counting, not the assertion.

So confusing is the nature of mathematical truth that there are statements in mathematics that seem true but have been neither proven nor disproven. One famous example is the Goldbach conjecture, which asserts that every even number is the sum of two primes. For example, 4 = 2 + 2; 6 = 3 + 3; 8 = 5 + 3. Such sums have been found for every even number up to 100 million. But nobody has established that the assertion, so simply stated, is actually true for *all* even numbers. Twentieth-century logic has gone further; it has shown that there are statements in mathematics which are indeed true but can *never* be proven.

So the minute we begin contemplating the notion of mathematical truth we run into a phantasmagorical swirl of issues philosophers have been pondering for millennia. Yet there math is, wherever we turn. We may not know what "pure thinking" is, but we do know it often comes in handy. "Whoever undertakes to set himself up as a judge in the field of Truth and Knowledge," Einstein warned, "is shipwrecked by the laughter of the gods."

MANY OF THESE problems disappear, though, if we alter our conception of those gods. There is a long tradition in mathematical thought, a tradition that goes back to its very origins in the surveying of land or the mapping of the heavens. In that tradition there is no problem at all in assessing mathematics's unreasonable effectiveness, or the nature of mathematical truth, because the world of mathematics is actually *more* real than the world, not less; it is primary, not secondary; it presents the essence of objects and their relations. This is not a mere metaphor. Plato maintained that individual objects only exist as imitations and reflections of a higher order of Forms. He believed the problem to be the inverse of the one we have posed; the challenge is not to explain how mathematics fits the world but how the world develops out of something resembling mathematics. Mathematics, in this view, is a divine language.

In many cultures' views of mathematics (and music), parallels exist with the world. These beliefs make Einstein's question about the connection between thinking and the world trivial: of course, thinking reflects the universe; it can do nothing else. We think the way the universe works. This was a strong tradition in Western philosophy of the last five hundred years as well. It was believed that when we discovered the properties of pure thinking, we would have a language that would automatically produce truths about real things—partly because the essence of real things was in the pure thinking. Leibniz, for one, hoped that philosophy could construct a language that would precisely reflect the universe: the only things expressible in it would also be true. The logician Gottlob Frege set out to define the laws of thought a century ago. Then Bertrand Russell sought to derive all mathematical truth out of logical truth: to prove that mathematics was founded in the laws of thought and that the laws of thought, in turn, reflected the truth. His program turned out to be unsuccessful, but the faith it represents is revealing.

These diverse efforts are rooted in the belief that the world can be described mathematically because its structure is mathematical. God, in this view, is a mathematician. In the eighteenth century—the same era that gave us the calculus and began to ask reflective questions about premises and connections between physical law and analytical thought—this belief was part of the development of modern physics. When searching for the rules that govern the revolutions of the planets, Kepler posited that each orbit literally created a pitch, a vibration in the heavens. Taken together, these pitches created a music of the spheres, a music of which his formulas were just the mathematical expression. In *Mystery of the Cosmos*, he wrote, "God himself was too kind to remain idle, and began to play the game of signatures, signifying his likeness into the world; therefore I chance to think that all nature and the graceful sky are symbolized in the art of geometry." The assertion is that God's likeness *is* the art of geometry, and that the art of geometry (and its reflection in music) is, in turn, the signature of God.

Galileo was still more explicit:

> Philosophy [Nature] is written in that great book which ever lies before our eyes. I mean the universe, but we cannot understand it if we do not first learn the language and grasp the symbols in which it is written. The book is written in the mathematical language, and the symbols are triangles, circles and other geometrical figures without whose help it is humanly impossible to comprehend a single word of it, and without which one wanders in vain through a dark labyrinth.

This unshakable faith in the figures of geometry and formulas of mathematics guided the research itself. For centuries the circle was considered the perfect geometrical form; and since anything so perfect would be reflected in the heavens, it was thought that the motion of the stars and planets had to be circular. So strong was this belief that, though the data did not support it, the theory of planetary motion became more and more convoluted, simply to preserve the exalted status of the circle: there were circles made within circles, epicycles upon epicycles, until by the time of Copernicus the revolutions of the planets were depicted with complicated looping circular forms, dizzyingly whirling around one another, in an attempt to match the increasingly sophisticated observations of heavenly bodies.

Kepler did not discard the mathematical faith in geometrical structure; he simply tried to find a different principle of order and shape. For example, "I undertake to prove that God, in creating the universe and regulating the order of the cosmos, had in view the five regular bodies of geometry as known since the days of Pythagoras and Plato, and that he has fixed according to those dimensions, the number of heavens, their proportions, and the relations of their movements." The radii of the six then-known planets, Kepler argued, should be the radii of spheres inscribed in the five regular solids: the cube, the tetrahedron, and so on.

Though the data finally did not support this elegant belief, the mathematical faith was still not discarded. In fact, Kepler spent decades of close observation, playing with numbers, determined to find in them a pattern that would be not just regular but also mathematically refined. The years of minute and laborious calculations, the obsessive attempts to fit numbers into predetermined schemes of form and beauty, finally resulted in a simple formula which spoke with the force of revelation. The planets, Kepler concluded, move in the shape of ellipses—curves that, like the circle, can be created by slicing a cone. In addition, the planets' orbits could be described according to surprisingly simple mathematical principles: the square of the period of revolution of a planet's orbit varies with the cube of its distance from the sun.

This is an extraordinary assertion; it almost seems invented with its formal precision. It is all the more remarkable that Kepler's discovery of this law came from the analysis of thousands of astronomical observations; it had nothing to do with a theory of gravity or motion. The law was not derived from some analogy (as Newton was to make) between the laws governing the motions of the heavenly bodies and the laws governing earthly ones. Kepler came upon his law because of his conviction that a mathematical relation had to govern the cosmos.

Here is Kepler's own reaction to his discovery, which is

strikingly similar in many ways to Bach's reactions to his compositions: "The wisdom of the Lord is infinite; so also are His glory and His power. Ye heavens, sing His praises! Sun, moon, and planets glorify Him in your ineffable language! Celestial harmonies, all ye who comprehend His marvelous works, praise Him. And thou, my soul, praise thy Creator! It is by Him and in Him that all exists. That which we know best is comprised in Him, as well as in our vain science. To Him be praise, honor and glory throughout eternity."

There is no question that without the faith that a description of planetary motion should be simple, elegant, and fundamental—that the universe itself is ordered in such a way as to be predictable and harmonious and accessible to the powers of mind—Kepler would not have arrived at his formula. Without an abiding faith, Kepler would have had no clear way to organize the data from the thousands of astronomical sightings. His was not just a faith in the order of the universe; it was a faith in the human power to discern that order. "All pure Ideas, or archetypal patterns of harmony," Kepler wrote, "are inherently present" in the minds of those apprehending them.

Newton went further. Not only was he, like Kepler, intent on finding simple laws governing the behavior of objects but he was intent on explaining why such laws might hold. This quest involved a still greater faith. So connected were theoretical notions and philosophical ones that Newton's description of God in the *Principia* makes the deity a sort of grand Fluxion, a divine representation of the principle underlying the invention of the calculus: "He endures forever and is everywhere present; and, by existing always and everywhere he constitutes duration and space. Since every particle of space is *always*, and every indivisible moment of duration is *everywhere*, certainly the Maker and Lord of all things cannot be *never* and *nowhere*."

What amazed Eugene Wigner and other twentieth-century physicists who have wondered at the results of such thinking is that this religious faith in a mathematical order was in fact rewarded. For example, Newton asserted that gravitational force varies inversely with the square of the distance between two bodies, basing his law on the measurements of speeds of falling bodies. But there is no way he could have measured the speed of falling bodies with a sufficient level of accuracy. His measurements did not really justify his conclusion. Wigner believes that Newton's formula could only have been verified experimentally in his time within a range of 4 percent. The fact that it was asserted nonetheless, without any expectation of error, and as a law governing all bodies in the cosmos—not just objects falling to earth—illustrates Newton's essential conviction that mathematics is the language of the universe, and that this language expresses relationships simply and elegantly. When measurements were made, they were considered more fallible than the mathematical relation posited between them. Newton had no convincing notion of how gravity works, yet he was able to describe its actions as a reflection of ratio.

But more astounding than the faith is its justification by results. Wigner has written that Newton's law of gravitation has turned out to be accurate to within less than a *ten thousandth* of a percent despite the crudity of his experimentation: "The mathematical formulation of the physicist's often crude experience leads in an uncanny number of cases to an amazingly accurate description of a large class of phenomena."

Things are not all that different in today's physics, which is more dependent on mathematics than it has ever been before, and more replete with unexpected applications of mathematical theories. Indeed, there are areas in physics which are first known only through mathematical theory. The faith is not just that mathematics represents the truth but that it represents the truth so much more faithfully than reason gives us any right to expect.

We are not about to explain this uncanny efficacy of mathematics, nor do we have a theory of theories that will precisely reveal the relationship between abstraction and application. But we can begin to understand why the powers of mathematics are not as implausible as they seem. When we began our exploration into music and mathematics, we thought about the nature of comparison—noting, for example, similarities between a raven and a writing desk. We had to define the characteristics we observed in each, consider what functions they serve, and decide whether there were any significant similarities. Finding links and connections between different objects is only one of the ways we make sense out of any stream of phenomena we are exposed to, passing by us like notes in a musical score or like seemingly random collisions of subatomic particles. We try to find repetitions, similarities, ratios. Our search for pattern means dividing phenomena, grouping their constituent elements into smaller categories, constructing "phrases," seeing whether, when we do group them together, we recognize one element's relation to another. We must also constantly decide what is important and what is negligible, which facts have significance and which do not. We do the same thing when we try to make sense of music.

Physics must abstract from the barrage of events in the world just such essential characteristics that might repeat, follow one another, or affect one another. If a physicist gazes at the heavens, it is to note repeated patterns in the positions of the planets or define (as did primitive peoples) the outlines of constellations; the patterns created today may be less animistic, but the motives are the same. The very notion of natural law upon which physics is based is an assertion that resemblances are to be found, that there is in each event not just the particular but an example of the universal. So physics, like music, takes as its project the recognition and articulation of form in the midst of phenomena. It attempts to find analogies. It should not be surprising then that mathematics is closely aligned with physics, since mathematics is in its essence (or as close to it as we are likely to come) the study of form and structure. Mathematics takes the techniques of abstraction and observation used in physics as simply one special case of a series of methods to be applied again and again. Mathematics, almost by definition, provides the method of physics.

That this is so may be more clear if we return to the notion of group touched on in earlier chapters. Suppose we are looking at the number system and want to figure out what happens when we divide numbers by 12. Sometimes we will get an integer as an answer, for instance, when the number is divisible by 12 (as in 12, 24, 36 . . .), sometimes we will get a remainder of 1 (as in 13, 25, 37 . . .), sometimes we will get a remainder of 2 (as in 14, 26, 38 . . .), and so on. In fact, dividing by the number 12 ends up giving us twelve classes of numbers depending on what sorts of remainders are left. We may call these ⓪, ①, ②, . . . ⑪ and show that every integer is a member of one of these classes. And we can, as in Chapter 2, develop an arithmetic based on these classes. For example, if we add a number that leaves a remainder of 1 (such as 25) to a number that leaves no remainder (such as 12), we get a number that leaves a remainder of 1 (such as 37), or, in the symbolism of our classes, ⓪ + ① = ①. Similarly, we can establish rules about the multiplication of these classes. They satisfy the requirements of what I described in Chapter 2 as a group.

The concept of group is quite general. It states that certain operations hold between members of the group and that certain rules of arithmetic hold. We can even generalize the example we have given. The circled numbers can represent anything at all as long as they follow the criteria we have set up to define a group of order 12 like the one just described. It turns out that this is the same structure that exists (with addition) in our telling of time: add two hours to eleven o'clock and you get one o'clock (② + ⑪ = ①). It is also the same structure that governs relations between the twelve notes of our musical scale:

add two semitones to the eleventh chromatic note of the scale and you reach the first semitone of the scale.

The same structure might appear elsewhere as well, in the positions, perhaps, of rotating crystals, or in the descriptions of a peculiar-looking surface. Thus, quite different systems are shown to have the same underlying structure. Mathematics has created a model that can be given an extraordinary number of interpretations. By defining the elements differently, we get a new meaning. We have thus defined a principle of organization that can be interpreted in unpredictable fashions and brought into the world in an astonishing number of ways. Mathematical systems provide the foundations for metaphors—forms to be carried from one place to another—giving us the same illumination as poetic metaphor does when it transports an object from one realm into another while leaving its structure intact. As Aristotle said, "A good metaphor implies the intuitive perception of similarity in dissimilar things."

Physics plays with the same principle: it seeks to find general laws by observing particular examples. The world is regarded as a single interpretation of a more abstract model. The very idea of natural law requires belief in a mathematical model lying underneath the surface phenomena of the world. The observable world is a metaphor for an underlying mathematical truth. A law, like Newton's law of gravitation, asserts relations without particulars; it presents an algebra of forces and interactions, a model that is interpreted whenever an apple falls to earth or the moon continues in orbit. Such a law does not explain, it describes; and each incident governed by such a law is an interpretation of the model the law creates. This is what gives physics a religious cast and closely connects it to the Platonism of early thinkers like Kepler and Newton. Modern physics seeks to find a coherent model that makes sense of all available and all possible information, providing us with a unified theory of the universe.

Such a theory was the quest of the early mathematicians as well. They were convinced that their work would reveal not just a form but Form itself, not just an instance of mathematical possibility but its governing principle. We know now that Form should be plural, not singular: mathematics can model other universes, not just our own; it might describe a world bearing little relation to ours; it might begin with hypotheses we sense and proceed to conclusions we cannot readily perceive. But we stringently apply the same techniques and expectations to each world, abstracting, yet again, from the results mathematics supplies, finding yet again similarity, connections, and proportions. We still treat mathematics to a great extent as did the Platonists. We may no longer appeal with Kepler to God, but when it comes to mathematics we tend to be transcendentalists.

The mystery may not be that mathematics has so many practical uses in the world but that our minds are so capable of comprehending them. There may be, as Kant in his way believed, a close connection between our experience of the world and the inner structure of the mind. The harmony found without may be a sign of a harmony within. As the mathematician Wolfgang Pauli once remarked, "The process of understanding in nature, together with the joy that man feels in understanding, i.e., in becoming acquainted with new knowledge, seems . . . to rest upon a correspondence, a coming into congruence of preexistent internal images of the human psyche with external objects and their behavior." Kepler put it more succinctly when he wrote of "Ideas, preexistent in the mind of God and imprinted accordingly upon the soul, as the image of God."

MUSIC DOES NOT offer a model of the universe in the way mathematics does or in the way the Pythagoreans believed, but it certainly offers an abstract form of *something*, which can be given meaning in any number of ways. The startling thing about music is not its "usefulness," which is practically nil, but its *power* in a culture's emotional life, social organization, religious ritual—something that we might call, metaphorically, music's truth. We have a sense, by now, of how music works and what it can suggest, but how does it take on such inarguable meaning for listeners? This is as difficult to comprehend as the corresponding question about mathematics. George Santayana wrote: "That the way in which idle sounds run together should matter so much is a mystery of the same order as the spirit's concern to keep a particular body alive, or to propagate its life."

We may begin to understand this mystery from its simplest manifestations. As we have seen, there is music that is meant for dance, or for marching, or for singing as the harvest is gathered or a home built or a fire kindled. Such functional music is deliberately constructed to have a specific meaning. But popular and folk music, however sophisticated it may become in its technical aspect, is limited in the kinds of musical worlds it can establish. The goal of such music is not to create an autonomous world but to relate to this one in a direct and unambiguous fashion. The rhythm is the rhythm of dance or festival; the melody is the melody of suggestion or intimation or appeal; the harmony is often simply organized tension and release, meant to be unambiguously suggestive and demonstrative. It is rare, in fact, to find an example of either folk or popular art that takes the rules of its art as guides for more extensive exploration or aims at the creation of fully imagined worlds with an intricate inner life. Its meaning is plain, its truths least profound.

The clarity of meanings in such music has provided the model for attempts to control music politically on the basis of the connections it makes: function and style are judged in an almost elemental fashion, as if music never dealt with anything more than simple identifications. Recall the tritone: it was called the *diabolus in musica* because of the kind of sound it would have introduced in a devotional setting; it was feared to invoke a world beyond the ordered system of rational composition, suggesting something threatening and unpredictable. Socrates, ostensibly devising a perfect state, argued that certain Greek musical modes should be forbidden completely. The Lydian mode must be dispensed with, Socrates noted in *The Republic*, because it would encourage "wailing and lamentations"; other modes, such as the Ionian, he considered too "slack" for use in training a group of warriors. Temperament and the temper of musical lines were clearly connected to temperament and the temper of the listener. Education in music, Socrates also insisted, must be carefully controlled because rhythm and harmony "insinuate themselves into the inmost part of the soul."

The attempts to legislate a socialist realist style in the late Soviet Union meant that the atonal and serial music of such composers as Webern and Schoenberg was forbidden, along with any other music that seemed too academic or troubling. These were technical regulations—seemingly abstract but with clear effects on sense and sensibility—and they were violated only in fear and trembling.

Even more easy to comprehend and control was music that set a text, or bore a program. It would be easy enough to find Berlioz's *Symphonie Fantastique*, with its erupting fantasy and passionate ejaculations, unsuited for a devotional church service, just as Beethoven's *Fidelio* with its vision of individual freedom was once thought unsuitable by Mao's China.

These brief examples of musical allusion generally involve treating some elementary aspect of the musical work as a representative image of another realm—the soul, the state. The mapping between music and other realms is usually trivial: taking the intervals as indicative of harmony between parts of the soul, or the presence of certain rhythms as models of bodily movements. We do not need a sophisticated vision of music to establish these sorts of relations.

The difficulties arise when we insist upon a more sophisticated and complicated view of music and its powers. And here the temptation is to accept yet another narrow explanation for music seeming to be "true." The usual meaning ascribed to music has long been a simple relationship to feeling. "By the musician's art," wrote Charles Avison, a British musicologist and organist in 1775, "we are by turns elated with joy, or sunk in pleasing sorrow, roused to courage, or quelled by grateful terrors, melted into pity, tenderness, and love, or transported to the regions of bliss, in an ecstasy of divine praise." C.P.E. Bach, one of Johann Sebastian's sons and one of the most influential thinkers and composers of the same period, wrote, "Since a musician cannot otherwise move people, but he be moved himself, so he must necessarily be able to induce in himself all those affects which he would arouse in his auditors; he conveys his feelings to them, and thus most readily moves them to sympathetic emotions."

These comments treat music as a concrete translation into sound of "emotion" and "feeling" and the act of listening to music as a sort of emotional sympathy. Avison described music's power to transport; Bach noted that the power comes from the musician himself being moved. These notions lasted until the modern era and nearly became a fetish during the period of High Romanticism. In 1854, for example, Eduard Hanslick wrote, "A performer is allowed to liberate whatever emotion sways him at the moment directly through his instrument and to breathe out in his performance the wild storming, the passionate flaming, the cheerful power and joy he is feeling within . . . In this situation, subjectivity emerges as immediately effective sounding in tones, not merely mutely forming in them." Music, in other words, is the direct expression of inner life.

These simple proclamations of how music works are also, of course, expectations of what music should be. The mercurial sentiments of C.P.E. Bach's music, its shifting affect and rococo melodic excursions, are well suited to a belief that invoking and expressing feelings is the purpose of music. Today we tend to assert that music's main goal is to provide pleasure in an accessible manner, a belief that is perfectly suited to the aesthetic of folk and pop: audience acclaim becomes the criterion by which music is judged.

But what of the formal, abstract understanding of music's inner workings we have been exploring? Or the notions of proportion and analogy we have discussed? Or the ways in which multiple processes take place creating elaborate sonic metaphors? How do the Western tradition's most systematically constructed works, its most intellectually sophisticated and most subtle creations, touch our spirits or souls or feelings? More is involved, obviously, than self-expression. As Susanne Langer has observed, "Sheer self-expression requires no artistic form."

What, then, is the function of "artistic form"? How is the exploration of a musical premise passionate? In what way does the construction of a composition affect us when we listen? What relations do the internal workings of a composition have with the effect of music, with notions of joy and horror and bliss and passion and knowledge? And are there any more profound connections that might, at least metaphorically, reveal truth?

BEGINNING WITH a musical phrase, we already have some notion of what such meanings might consist of. We know that a phrase presents a series of tensions and relaxations; it implies a certain sort of movement, creates expectations it might disappoint or might fulfill. A phrase is, as we have seen, a gesture; it has the precision of movement and character that we sense in a shrug of the shoulders or a wave of the hand. A gesture can be approximated by language and by explanation, but these would be only approximations, translations, not substitutions. A gesture is complete: it can stand on its own; it possesses a beginning, middle, and end; and it makes a statement in space and through time. It is meant to be an atom of meaning, a movement which, if interrupted, would hardly convey its sense at all.

The musical phrase is, as we have seen, just this sort of object. It is not simply an expression of an emotion or feeling or thought; it is not a translation of something else. It is self-contained but also full of implication and possibility. Consider, for example, the familiar motive of Beethoven's Fifth Symphony. Its most immediate impact is rhythmic: it begins with three eighth notes, followed by a long sustained tone. Properly played, this motive, as short as it is, is a gesture. In the popular image it is a variety of Morse code (the dot-dot-dot-dash code for the letter *V*, which served as a convenient representation for Allied victory during the Second World War). But that gesture is physically stiff, stompish, a tat-tat-tat-rat without shape. It should be, instead, supple, tense, even mysterious. The opening beat of the motive—the beginning of the symphony—is not that first proclamatory eighth note but a rest that precedes it, a silence equal in length to one of the eighth notes. That rest is peculiar because it opens the phrase; it is still more peculiar because it is the opening of a measure. Like the first beat of every measure, it is *accented*.

What is an accented silence? How does it begin a phrase? It can only be felt in the way the sounded notes are played. If we land on the first note as if nothing preceded it, it will simply stand starkly in musical space, an abrupt stroke. Instead, Beethoven seems to have intended that first sound to begin as if it grew out of the accented silence, as if there were an intake of breath, or a snap of the wrist, before it is heard. Without being accented itself, the tone must draw attention to itself and immediately pass attention to the next note, which should be more strongly accented. So the opening note is weak but is given a sort of latent power by the preceding silence; the second note is stronger but is heard as growing out of the first. When the final note of the theme arrives, it is the apotheosis of accent; it is a suspension of time and breath, the accented rest of the beginning transformed into sustained sound. This is not Morse code; it is breath and gesture.

There are many ways of shaping that motive, but if they don't make sense of that opening silence, and the fermata, the extended suspension, of the final tone, they miss the character of the symphony itself and its extended play with accent and extension. We are *not* free to play this motive any way we like; it is not languid or impressionistic; it cannot be successfully played without accent, or without attention to rhythmic relations. But there are any number of ways in which it *can* be played; it is beyond simple characterization. We might call it, as Beethoven's amanuensis once insisted Beethoven did, the sound of Fate knocking at the door. But what this motive *feels* like, what it suggests to a listener, is a matter of breath and pulse, pressure and extension. Its impact is immediate, available to a listener without reflection, because the reflection has already been done by the composer and the performer. Relations already have been turned from paper-bound facts about accent into audible facts of relation and movement and smoothness; thus, musical space is created.

In creating a gesture of this shape, Beethoven expected his listeners to understand it in a particular way; the gesture had a context. Its physical setting was the concert hall; its musical setting was the symphonic style developed by Haydn and Mozart. And within that framework, Beethoven's theme had another gestural character: an element of unsettling shock or surprise. His listener would likely have expected a theme lasting eight bars, with a melodic, possibly even conversational character. Beethoven does eventually provide that, but his theme is little more than a fragment. It would have startled the listener of 1808.

Our response to such musical gestures is worth some reflection. Understanding the nature of a gesture in music requires understanding the context for gesture, both within and without the work. A shrug of the shoulders may be good-natured bonhomie in one context and an insult in another. Thus, when we learn to hear a composition, when we learn a style or appreciate a different culture's music we learn how to hear its gestures. Leonard B. Meyer, in *Emotion and Meaning*, his compelling psychological analysis of music, has suggested that, in the West, grief is communicated with a particular kind of behavior: limited motor behavior, "sorrowful" countenance, weeping. This is precisely the sort of behavior modeled in some Western music—Bach's and Handel's arias of weeping and sorrow try to imitate physical behavior in their sighs and sobs. The reason music requires intensive study on the part of the listener even as it seems so immediately understood is that we need extensive experience to comprehend the gestural language of any composition, even one as familiar as Beethoven's Fifth Symphony.

Gesture is at once cultural, physical, and formal. Gesture is cultural because it is part of a network of nuanced signs and references, some belonging to an age, others to a place. It is physical because the differences in pitch and accent and harmony reflect internal feelings of movement and pulse and pressure and expectation. Our concert halls discourage all physical movement, though, because when the physical model becomes primary we miss the more profound formal gesture. In the Beethoven theme the senses of extension in the final tone, of pulse in the earlier ones, in the differences in attack required by accents, are matters of how one musical element relates to another. They are, as we have suggested, musically abstract relations; they compose the formal gesture of the theme.

The formal gesture might be considered an abstraction from the physical one that, like some kinds of mathematical movement from the concrete to the abstract, soon becomes independent of it. Such abstractions are what make it possible for music to construct its own world based on its own premises, without constant reference to the physical world. They allow an almost formal unfolding, a series of events that are abstractly ordered, yet can be mapped into a wide variety of physical and emotional and mental worlds. Each mapping creates different meanings just as does the mapping that interprets a group of order 12, or the mapping that provides instances of a natural law.

When we hear the theme of Beethoven's Fifth Symphony, we also attend, even without being aware of it, to the formal gesture, the abstract phrase. We respond, for example, to its uncertain musical center. How, in the repeated G and sustained Eb, are we to guess what the tonic tone is? We don't have any clear way to "place" these notes; where are they going? This sense of uncertainty is at odds with the certainty of their statement; that uncertainty increases when the same pattern appears just after the fermata on an F and a D. The sense of anxiety remains even as the music continues. The pulsing of the theme, with its characteristic rhythm (it hardly matters which notes fill that rhythmic form), contributes to the nearly Baroque suspense of this movement. It is almost brutishly simple in its repetition of material and its manipulation of our expectations.

Yet we already can see from this description that no matter how much we consider the gesture itself and the fields of force it creates within the composition, we can never speak about it in a *purely* formal way. In music, we must always interpret unfolding form, mapping continuously into meanings, creating metaphors. We must even mix metaphors when we speak about music, peppering our vocabulary with highly charged words like *single-mindedness, anxiety, suspense, ambiguity, expectation.* Metaphor is unavoidable. The very act of describing a formally organized system, of giving it notation, or sound, or connecting its elements to sensation creates a metaphor; it says that one thing is another, that a raven is a writing desk or that this music is "about" a relationship between repetition and transformation.

Even when we go through a composition applying harmonic labels to every chord, as in traditional musical analysis, we are not merely articulating form; we are defining the composition metaphorically—mapping it into a universe composed of named harmonies and chordal relations. When Schenker searched the composition for fundamental lines and descending scalar figures, he did the same. Once we describe music, we create constellations, applying to the musical realm the sort of attention physics applies to the physical one. The music "in itself" is the abstract model whose essence defies even a purely formal analysis.

So the interpretation of music—or its description in words—is like the interpretation of a mathematical model; it is a transformation, a mapping from one system into another, from the "score" to another structural representation of it, an interpretation of its significant elements and relations. These metaphors can be as precise as our most sophisticated academic vocabulary or as suggestive as a poet's image. E.T.A. Hoffmann was one of the first serious listeners to Beethoven's Fifth Symphony. Here's how he chose to describe what he heard:

> The instrumental music of Beethoven open[s] the realm of the colossal and the immeasurable for us. Radiant beams shoot through the deep night of this region, and we become aware of gigantic shadows which, rocking back and forth, close in on us and destroy all within us except the pain of endless longing ... Only through this pain, which, while consuming but not destroying love, hope, and joy, tries to burst our breast with a full-voiced general cry from all the passions, do we live on and are captivated beholders of the spirits.

Such a description seems entirely opposite to the forms we have been exploring. What have "radiant beams" or "gigantic shadows" or "pain" or "spirits" to do with Beethoven's motifs or his musical style? Hoffmann noted that from the first twenty-one measures of the piece, with their fermatas and repetitions, "a presentiment of the unknown, of the mysterious, is instilled in the listener." This is not just airy hand waving; it is a valid way of explaining the music: we can translate this statement, undo its metaphor. The "presentiment," for example, is in the floating nonkey of the opening, the insistence of its rhythms, the way overlapping motifs seem to nip at each other's pronouncements. More generally, we might identify the "pain" of Beethoven's music with the almost obsessive repetition of themes that are gradually altered under this insistence. We might link "radiant beams" with the sort of comprehension that comes when we sense a theme return to the tonic in its simplicity after having been subjected to the most exotic adventures.

Of course, these relations are not as clearly defined as in mathematics, nor are they as rigorous, but they are unmistakable. The most abstract music, if it is heard as music rather than as a collection of sounds, if it is heard as an order, a construction, which is meant to "make sense" according to its premises and ambitions, is already apprehended by the listener as programmatic or emblematic or metaphoric. A program is simply a particular interpretation of a music's workings. It can be imagistic, as in Berlioz's opium dream in the *Symphonie Fantastique* or in the image of a sunken cathedral rising from the ocean in a Debussy prelude. Or the program can be quite technical, as in an attempt to hear the first movement of Beethoven's Fifth Symphony as an analytical meditation on the opening four bars.

In fact, every considered performance of a composition turns music into program music, creates constellations and narratives, references and transformations, all selected out of the score's written universe. One conductor will make the first movement of the Fifth a programmatic work about the rhythmic gesture of the motive; another will find a program in the ways the two statements of the motive that begin the movement relate to each other. Earlier in this book, I turned the *Appassionata* Sonata into a disquisition on C and D♭. Music may possess what Thomas Mann called "cow warmth," but it also has the power to prompt interpretation. It presents an abstract world whose meaning is paradoxically both determined and open.

This paradox at the heart of music is why it is possible to speak with such ease about music in spiritual or literary or intellectual or technical terms. Once we accept the metaphorical basis of the interpretive connection, we can concoct any number of programmatic dramas—many with profound implications. If, for example, we hear in the opening measures of Beethoven's Fifth something that resembles the insistent voice of the individual will, then we can begin to interpret the symphony itself in terms of its approach to the individual and the community. The orchestra seems an embodiment of the individual, its body and mind, in which all musical events are internal, personal, even as they are made public and demonstrative. We can make association with individuality still more convincing by describing Beethoven's almost obsessive control of every aspect of musical sound. As Theodor Adorno, the twentieth century's most idiosyncratic Marxist aesthetician, wrote, "If we listen to Beethoven and do not hear anything of the revolutionary bourgeoisie—not the echo of its slogans, but rather the need to realize them, the cry for that totality in which reason and freedom are to have their warrant—we understand Beethoven no better than does one who cannot follow the purely musical content of his pieces." Adorno was suggesting that in the way Beethoven employed musical material we can find a new attitude toward musical convention and style, a use of repetition, and a manipulation of thematic material that is individual and willful but is also controlled and calculated. In Beethoven's music, the idiosyncratic, controlling will seeks to be part of a coherent whole, a public in whose name it speaks.

Composers too can make plain their intentions. When Bach inscribed his scores "J.J." for "Jesu, juva" ("Help me Jesus") or "S.D.G." for "Soli Deo Gloria" ("To God alone the glory"), he was invoking the traditional metaphorical link between music, God, and the soul. But he made these metaphors part of the substance of his music as well. In the B Minor Mass, the incarnation of the Holy Spirit is illustrated by the sighing violins descending into the low-pitched earthly realm of the bass line. Much of music, for Bach and for many of his predecessors in the Church, was explicitly religious: the metaphor for musical construction was the composition of the spiritual life. Studies of the technical aspects of counterpoint and harmony were in service to another order: S.D.G.

But Bach's music allows us to embrace other interpretations as well. His fugues are quite different in their effect and character from anything that came before or after. In a Bach fugue each voice is free from the others, independent and fantastical; yet it is contained in a structure in which the degree of repetition and imitation is immense. The fugue sets up a community of like minds and distinct parts, quite different from the polyphony that preceded it, which created textures rather than lines of individual character. The Bach fugue is quite different too from the fugues of other Baroque composers, which often engage in repeated imitation and leave it at that. Bach's fugues, though written in a religious period and by a religious man, contain premonitions of the secular Enlightenment; their themes almost become objects of rational knowledge, their every gesture and interval subject to study and illumination.

This speculation interprets Bach's music using metaphors which can then be subject to refinement, clarification, even correction. Metaphor is not an indulgence. It is a way of reaching essence. Metaphor helps us to understand why some compositions affect us in one way and others do not; it reveals what we mean by the content or meaning of music. It even helps us understand the music's public meanings, its role in the world of politics. Whether a piece is intended for public or private performance, whether it is written to court an audience or challenge it; whether it is written out of obligation or inspiration; whether it is written for money or glory or glamour or other reasons; whether it is democratic in temperament or authoritarian in spirit—all these distinctions affect how we understand particular compositions, and all can be "coded" in various ways in the abstract system of relations that music establishes.

We can even step further back from the entire system of Western music and derive meanings from the series of formal and informal styles that evolved over a millennium. Max Weber, for example, noted that the development of Western music followed the course of "rationalization" in society—the displacement of religion by civil authority, the increasingly intricate structural relations between social organizations, the systematization of knowledge itself. Music, he argued, passed from religion to science and secular life, from manipulation of repeated pattern to the exploration of hierarchy and structure, from regulatory boundaries on harmonies and intervals to attempts to create a form of abstract knowledge about combinations of sounds. Music might be considered, in this light, as a counterpart of Western science.

THE EXPRESSIVE possibilities of music lie in connections made between formal systems and their interpretations as physical sensations, emotional affects, mental ideas, poetic images. But the freedom and inherent ambiguity of impressionistic interpretation have often scared off critics and academic theorists. It has been easier to symbolize and analyze internal language, to name harmonies and operations on sets—essential activities, to be sure— than to engage in the sorts of critical considerations we demand from other artistic discussions, mixed metaphors and all.

There are risks in the other direction as well, as we can tell from the increasingly fanciful interpretations now being made (even our own is not without risks). But we can know one thing clearly: any musical system, like any social system, creates a set of manners and traditions, laws and customs. It subjects the events in its realm to analysis and reflection. It is also fundamentally involved in establishing an order, which, as we have seen, depends upon the creation of boundaries between the forbidden and the permitted, the dissonant and the consonant. There will always lie regions beyond the bounds of music's realm—areas of noise and silence—but without those regions lying beyond its reach, the order of music could not possibly establish itself. Music even takes this into account: dissonance is the presence of noise in musical structure, noise controlled and put to use, turned, paradoxically, into music.

The descriptions of the processes used to create order and demarcate noise might seem abstract, riddled with generalities, but the musical order thus created is also a creator of order in its listeners: the audience is made into a society by the music. There are even aspects of musical composition that can be politically subversive. John Cage, for example, polemicized against musical order altogether; he advocated the notion that all sound, even randomly arrayed, can be called music; his stance was of a piece with his anarchistic social polemics.

Music is so supple a bearer of meanings that it can, like the most subtle uses of literary language, suggest attitudes not easily articulated, at times asserting contradictory poses simultaneously. The case of Dmitri Shostakovich, for example, is instructive. On two occasions he nearly lost his freedom (and, of course, his life): in the 1930s he was attacked by Stalin when he celebrated the unbounded passions of sex and murder in his opera *Lady Macbeth of Mtsensk*; in the late 1940s he was accused of being "pessimistic" and "formalistic." The composer responded in both cases by seeming to "confess" and then "reform." Shostakovich became one of those rare figures in music history: he could be considered a "court" composer whose job was to turn out music that would keep the Party happy; he wrote music to order—according to subject (the greatness of the Russian people) and style (always tonal, never overly jarring). Moreover, his music had to affirm the existing social order, not disrupt it.

Most purely affirmative music risks losing itself: it is partly the presence of disorder and risk that gives music its power (the devil makes Faust a plausibly musical subject). Yet Shostakovich's music did not fall prey to the dangers of affirmation, partly because he had the talent and the cunning to use it as a mask: properly played, the music turns to sarcasm, a voice emerging from a condemned universe. In the West, most critics regarded Shostakovich as a craven Party composer, skillful but overblown, dutifully toeing the line. Then, with the publication of his memoirs in 1979, his attitudes of dismay, despair, opposition, and horror toward Stalin came out, and they were confirmed by the testimony of his family and friends. Suddenly, his music took on new meaning; its strange textures were now heard as ironic. Instead of aiming to create a single beatific community, united in the splendor of self-congratulation, Shostakovich's music splinters that community, attacks and divides, even turns against itself. Yet it always preserves the possibility of being heard in completely the opposite manner.

The social meanings of music—and their relation to compositional techniques and styles—merit their own study. For our purposes it is enough that music can legitimately be regarded as a model—both concrete and ambiguous—of the structure and order of society itself. Performed sound reflects back to society the shape it has received. Music's creation may be private, its workings abstract, but its impact and cultural position are inescapably social. It is no accident, for example, that the music of our contemporary culture is fundamentally the music of the nineteenth century, which was the first music written for a paying public, for massive orchestras, for grand secular halls. It was the first music written for and about the emerging bourgeoisie, which is why middle-class listeners today are still absorbed in its message and meaning. Our social order is inscribed in the art's substance.

PART OF THE SOCIAL FUNCTION of music—the most mysterious part—is its role in ritual; this is also a function that reveals some of music's most important meanings. Music, like ritual, involves the transformation of a community of belief—the community of listeners who are willing to accept its premises as if on faith and follow them into uncharted regions. Music is performed by trained initiates for such a community. One of its purposes is to transform those who gather in silence to await revelations. Music, like ritual, discards our ordinary experience of time, sometimes suspending it altogether, always changing it. And music, like ritual, is fundamentally written to be repeated—repeated in such a way that it becomes more powerful over time rather than less. What is unclear is just what sort of ritual music is.

A religious ritual is a repeated act that invokes through its details the workings of other realms—the worlds of cosmic forces, or human desires, or magical beings. For example, in Judaism, the songs sung when welcoming the Sabbath make references to a mystical union between Israel and the Godhead, a union ending centuries of both physical and metaphysical exile; the Sabbath is addressed as Israel's bride. The welcoming of the Sabbath enacts a transcendental event, a marriage between God's people and the Godhead that also ends a rift in the universe. The Catholic Mass too is an earthly enactment, using objects that are more than mere symbols: the wine *is* the blood of Christ; the Eucharist *is* an event in the lives of the Savior and the participant. Ritual—like the mappings we have been examining—takes metaphor seriously; we might say it looks at metaphors as if from God's eyes, as equivalent to the objects we say they represent.

Ritual is thus a form of profound mapping. And, as we hinted when we began, it is another realm in which music and mathematics seem essential. Mathematical connections to religion and ritual are plain. First, math is used as a tool; ritual requires precision and measurement, proportion and dimension. In fact, as we have seen, mathematics in both Greece and Egypt developed along with the priesthood and the need for structured ritual and measurement. But mathematics also creates its own models of universal order. There are steps that must be taken to do recognizable mathematics. They may vary from period to period, but without exception a nearly ritualized sequence of reasoning is used as the touchstone for mathematical proof; math is surely not only ritual, but at the heart of any mathematical style, ritual is present.

The ritualistic nature of mathematical thinking is most clearly asserted by mathematicians themselves, who comprehend how different are the processes of exploration we have sketched and their formal presentation. The great number theoretician G. H. Hardy wrote, "If we were to push it to its extreme, we should be led to rather a paradoxical conclusion: that there is, strictly, no such thing as mathematical proof; that we can, in the last analysis, do nothing but point; that proofs are . . . gas, rhetorical flourishes designed to affect psychology." It may be that the proof of a mathematical idea is like a musical performance, an interpretation of the content presented, and an attempt to present it as immutable and authoritative.

In a book of philosophical dialogues, the twentieth-century philosopher Imre Lakatos attempted to show how a simple proof is not simple at all but is a "thought-experiment"—an attempt to create, through definition and reasoning, a mathematical construction that serves a very particular function. Its statements may present themselves as absolute, but that is so mathematical proof will be unquestioned and taken to be revelation. In this radical view, mathematics is not a window on immutable Truth, but the creation of Reason's rituals, laying out an order and the steps by which an initiate may pass from darkness into understanding.

Music is, of course, experienced in an even more ritualistic fashion. We have seen how music's meanings are metaphorical—a phrase from Beethoven being linked to a particular way of thinking and feeling; music's power arises partly from the fact that these metaphors are *not* experienced as metaphors. Music has an immediacy and a transparency that is seemingly unconstrained; there is no distance between the sound and our interpretation. Music seems to gaze within us with God's eyes; we don't even notice the metaphors we make. This is why music's ritualistic powers have long been the subject of myth, the sounds touching regions beyond the reach of ordinary language or action: Orpheus taming animals with the sound of his lyre, David taming Saul's madness with his psalms, Pythagoras exorcising the demons of the possessed. Even in its most aestheticized forms, music creates a structured order, which—through gestures and mappings we have only begun to comprehend—seems to touch the life of the mind and heart, and even the heavenly spheres.

But in our secular musical culture, the ritualistic aspect of music is almost invisible. Unlike the music of the Renaissance or contemporary folk music or some music of non-Western cultures, contemporary Western art music is not often linked to ritual practice. Yet ritualistic elements persist. Their character depends intimately upon the kinds of worlds being invoked in the music and the kinds of communities being created by it. The particular ritual and its significance are different in Sri Lanka and New York; different in Bayreuth and Salzburg; different in twentieth-century Milan and eighteenth-century Paris. One way of comprehending the state of Western musical art during the last two hundred years is to stand back from the particularities, the variations in style and setting, and listen for something more fundamental about the assumptions we make about music and its functions.

Claude Lévi-Strauss provides a cue. He argued that though this period has seen the development of music as an art that was meant to be autonomous, independent of origins and function, engaged in the pursuit of beauty or truth or aesthetic discovery, it was precisely during this period—particularly during the nineteenth century—that music embraced the ancient function of myth. The central works of the Western repertory, I think, may be understood as tales which, in their highly metaphorical fashion, outline our society's origins, detail its passion, and trace the adventures of its heroes. That is what makes the work of Wagner so powerful, and that of Beethoven so central to our understanding of music. They define the aesthetic and cultural and religious foundations of modern society; they also celebrate the efforts of the independent will, and the possibilities of transformation through understanding.

If music has assumed the function of myth in Western society, then the concert has been a ritualized means of presenting the myth. The traditional concert typically began with a showy overture, proceeded to a nineteenth-century concerto with a renowned soloist, and, following a twenty-minute intermission, concluded with a "serious" nineteenth-century symphony. This progression suggested a movement from unrestrained and playful ego in the overture through confrontation and dialogue—the hallmarks of the concerto—to the creation of a social enterprise in which individual musical voices were bound in a grand, highly ordered community.

The very structure of the concert enacted a perennial theme of Romantic music: the individual passes through introspection, dream, and reflection and emerges transformed, embraced by a social order. The underlying belief was in the renewability of bourgeois society itself—and the concert hall was the temple in which this renewal is enacted. The music of the nineteenth century—the first music which was constructed in a narrative, novelistic style—has been repeated ritualistically in our secular temples, recounting our social origins in epic tales of extraordinary drama.

Exploring these musical meanings brings us to regions outside the scope of our inquiry: the shifts in the form of the concert; the influence of recordings on music's meanings and function; the power of the pop music industry and its connection to the more abstract themes we have been discussing; the trends in new music and the social forces being invoked; the evolution of the virtuoso. Our entire tradition is now at a point of transition. But at this late point in our own journey, we might see that, for all its variation and nuance, up until this moment in Western music its development corresponds roughly to the enterprise of science itself—science not as in the popular and poetic image of murdering to dissect, or as hypertrophic rationality—but as we have glimpsed its workings in mathematics: the attempt to make sense of the world by thinking about it in a particular way. The incorporation of the unknown and the urge to expand ever further the boundaries of the known and the powers of reason—these are its goals.

But if music has developed along these lines, with its explorations presenting abstract models which we then map into our lives, we can treat music as if it really were *about* the world, as if it really were concerned with the inner life of the mind, the shape of society, our conceptions of the universe. In this light, a composition is fated to be an unverifiable theory of the world, experienced rather than expressed.

Once more, we are returned to the imprecision of metaphor. The finding of music's "truth" is not easy. We are unable to keep the various systems we create free from one another's competing demands. All musical metaphors become hopelessly mixed as we attempt to find music's "meanings"; we speak of harmony and spirit, of rhythms and alienated consciousness, of narrative voice and musical influence. Our greatest risk is that our metaphorical interpretations will be willful, arbitrary, unenlightened, that connections will be made of trivial importance.

For example, one scholar, Arnold Schering, intent on establishing interpretive models, set up literary parallels for most of Beethoven's piano sonatas, supposedly derived from comments about their meanings and programs made by Beethoven himself. Thus Schering believed that Op. 27, No. 2, was based on *King Lear*, Op. 53 on Book 23 of Homer's *Odyssey*, Op. 57 on *Macbeth*. He even included German dialogue from some of the works he mentioned, dialogue he thought was scanned by the music. As Eric Blom has commented, "One can only wonder, by the time one comes to the last Piano Sonata, Op. 111, for which 'Henry VIII' has to do service, that the theme with four variations and coda of the second movement are not made to stand for the much-married monarch's six wives."

How can we possibly prevent such strained interpretation, given music's inherent freedom, its openness to our hermeneutic energies? Critics often find refuge in a set vocabulary, calling the balls and strikes of a performance culture (I confess I often find refuge there myself). Analysts work out theories of order. Academic theorists often take texts as canvases for painting pictures of various political or literary models. But a score's inherent abstraction is also its strength: it embodies numerous interpretive possibilities yet is more rigorously precise in its application to each one. We can find many interpretations, but the constraints on each are palpable.

How do we, then, in critical interpretations of scores or in evaluations of performances, determine which are valid and which invalid? Which violate the law of the score, and which elaborate upon it? Once more we must turn to metaphor.

As in a mathematical theory, we must ask that an interpretation establish a certain consistency. In mathematics that means it must not produce contradictory statements. Music, of course, prompts contradictory sensations, but since we listen to it in a state of poised contemplation, we require of a composition that it permit the construction of consistent metaphors without undermining them. We ask that it have a style and rhetoric and formal character that are open to consistent interpretation.

For example, in a piece in the Classical style, there is a kind of rhetoric of statement and counterstatement, an expectation that the piece follow a narrative. If the middle of a Classical sonata were to be interrupted by the sounds of Debussyan Impressionism, we would sense something inconsistent, a penetration by one universe into another. Of course, it can be one aspect of style to create such disruptions, to quote and invoke various themes and styles, such as in Luciano Berio's *Sinfonia*, which invokes varied stylistic voices. It is also possible for the premises of a composition to undercut themselves deliberately, so that the "repertory" nature of a work is dissolved. This has been the strenuous activity of the avant-garde for nearly the whole of the twentieth century and has become a defining characteristic of post-modernism. Consistency, in other words, is a matter of purpose and execution in music; its precise character depends upon the style of the composition. Even its apparent absence can become a type of consistency.

We also ask for something like "completeness." In mathematics, completeness means that all true statements are produced by a mathematical system. In geometry, for example, if two triangles have respective sides of the same length, then they should have the same angles as well; if our system is "complete," we should be able to prove that proposition to be either true or false. In music, completeness would mean—metaphorically—that the composition "fills" our listening universe, so that we do not expect anything other than what it gives us, that nothing other than the universe of the composition seems relevant to us; it fulfills our expectations. If we hear, for example, some work for the Baroque French court, like Jean-Baptiste Lully's opera *Atys*, it creates a world so completely hypnotic that it almost banishes any other.

These notions are admittedly themselves quite metaphorical, and I would never suggest they can be applied with the philosophical rigor of their mathematical counterparts. But the musical realm is the realm of suggestive metaphor rather than exact declaration. We know when we talk about "truth" in mathematics that we are talking about something that can be fairly well defined. But there is something "true" about music as well, something that concerns the kinds of thinking and feeling I have been describing. When we say something musical "rings false," when we say music "lies," we know what we are talking about. Musical truth involves something about the ways a composition is consistent, complete, and open to mappings in our various worlds of sense and thought. It is easier to specify the opposite: we think of music as false when it violates its own premises.

When we listen to Beethoven's Fifth Symphony, we feel that each moment is earned by the work; it belongs; the sensations we experience of suspense and harmony, tension and release may surprise us, but they do not jar us out of Beethoven's musical universe. When we listen to a sonata by Czerny, we are less certain. The music uses certain gestures as if they were ready-made, without justification; it proceeds glibly, full of casual ease even in the supposedly impassioned moments; and when the music signifies tension or grandeur in volume or insistence, it seems unearned, as if a formula were being applied. When the musical model is not seamless, when it is flawed and full of forced and artificial mappings, it resembles a failed attempt at a mathematical proof, or an early draft of a polished one. Its scaffolding is in no way disguised. Post-modern attempts to undo the dominant aesthetic demands of consistency and completeness thrive on the display of scaffolding.

The study of the Western tradition's greatest works offers

a different kind of vision. In the truth we establish for music—in the feelings and thoughts and meanings it creates, in the interpretations it inspires, in the mappings we make from it to our experience, even in the process of coming to understand how something as ineffable as music can resemble something as concrete as mathematics—in that truth we begin to discover something else, something we can just begin to glimpse in the next chapter, as our journey draws to a close.

VI

CHORALE:
THE TEXTURE OF THOUGHT

All of this is a prelude to the song itself which must be learned.

PLATO

O UR ATTEMPT TO COMPREHEND MUSIC AND MATHEMATICS, TO understand their workings and their purposes, is, we now can see, a model for our coming to *know* at all—a model for our education, for the ways we make distinctions and connections. Recall, once more: we begin with objects that look dissimilar. We compare, find patterns, analogies with what we already know. We step back and create abstractions, laws, systems, using transformations, mappings, and metaphors. This is how mathematics grows increasingly abstract and powerful; it is how music obtains much of its power, with grand structures growing out of small details. This form of comprehension underlies much of Western thought. We pursue knowledge that is universal in its perspective but its powers are grounded in the particular. We use principles that are shared but reveal details that are distinct.

And now, having defined—at least abstractly—a mapping between mathematics and music, we must step back once again and look at the process by which we have come to this place.

The final portion of our journey might seem strange, for we are going to deal only with the form of our understanding rather than its content. We are not going to examine music or mathematics at all, but only the ways in which we come to know them. That is, we are going to continue using metaphor.

The metaphor was given to us by Plato, when he attempted to reveal the nature of learning and understanding in his *Republic*. That great dialogue is ostensibly an attempt by Socrates and a gathering of friends and students to determine the nature of justice. Like so much else in this surprising text, however, the real subject is not what it appears.

At the heart of the book is Socrates' famous account of prisoners held in a cave. They are immobile, bound their whole lives to stare ahead at a wall of the cave on which shadows are cast by huge puppets and statues and artifacts being paraded between the prisoners and a flickering fire. The prisoners see neither the fire nor the puppets, only the shadows. Thus, the prisoners believe the universe of shadow to be the true world; they could perhaps even devise whole theories of motion and shape based on this universe; they might note repetitions and variations; they might even feel satisfied with the explanations, theories, and analyses that they develop.

If one of the prisoners were freed and turned to look at the light, he would at first be blinded, unable to make out the objects which until then had composed his familiar universe. He would not even be able to recognize the puppets as objects creating the shadows, and would first consider them as *less* real than their images on the wall. To understand the relationship between the shadow universe and this new world, he would first have to realize that there was a similarity between the two worlds he was observing—between the objects being moved and their shadows on the cave walls. Then he would have to realize that one was an image of the other. He would have to recognize that what he had once considered objects were mere reflections of another realm. The world of the puppets would not invalidate his earlier knowledge of the shadows: he would still be able to predict the movement of shadows and describe the laws governing their relationships. But these laws would become special cases of a more abstract universe; new laws would be devised to describe the relationships of puppets; they would absorb and contain the earlier theories of shadows.

If the prisoner were then brought up out of the cave, into daylight, he would have the same reaction again—of being blinded and confused, until he could make out the objects of which those puppets were only images. And again he would find that the ways the puppets were moved were really only special cases of much more general laws governing the objects of the world outside the cave. Finally, Socrates asserts, the prisoner would turn to beholding "the things in heaven and the heaven itself," with the sun as source of all illumination, all shadows, indeed the source of life. Theories of this world would contain the others. Comprehending the motion of the stars would reveal the principles that govern the shadows, the puppets, and the objects on earth itself.

If he were then to return to the cave, to the world in which shadows are the only objects, the prisoner would again be blinded and confused. He has, after all, reached a high level of abstract understanding, but that abstraction also may cloud the differences between levels of perception; the shadows and the puppets and the objects all follow the same principles, so how are they different? The first steps of descent might lead to confusion, bewilderment. Moreover, his account of the upper world would be thoroughly rejected by others. But his knowledge would also give him additional insight; he would begin to make more subtle distinctions between worlds and come to know their profound differences. He might then feel a duty to bring his compatriots into the light.

This story is Plato's model of how we obtain knowledge.

Our eyes are blinded by every move from one world to another, unable to distinguish the real object from its image. The struggle upward first involves a sort of *backward* thinking: the eyes meet each new object not as the more "objective" one but as if it were an image or transformation of the shadowed one; the resemblance of puppets to shadows might first be taken as imitation. But we then learn that the reverse is true, that puppets account for shadows in ways shadows cannot account for puppets. Each "real" and "final" object is just an image of an object found elsewhere. We then make distinctions and note similarities within each world and find their parallel similarities and laws in higher worlds, leading finally to a metaphysical realm governed by the heavens, which is the most abstract, the most encompassing, accounting for all we have seen and showing us the latent relations and structures in the worlds we have explored. We then return with our knowledge, wishing to share it with others. Education, says Plato, is the "art of this turning around."

Plato's dialogue dramatizes this art of turning. Socrates, for example, describes how to educate the citizens of an ideal city, arguing they must be told lies to keep them noble and good; all poetry and music they hear must be limited to set good examples. No bad things should be said about the gods lest people imitate them. There are no dissonances in this education; the well-educated man must be "perfectly musical and well harmonized."

But Socrates knew just how tenuous such an education is since he concedes that these very citizens could easily be undone by "wizardry or force." He makes it clear, without ever saying so, that this is an education based not on truth but on imitation; not on conviction or argument but on indoctrination. Later on, Socrates even proposes adding to his education of this mythic city a "noble lie." The citizens will be informed—ridiculously— that they actually received their education *underground*. With great cunning, this bizarre lie adds to the swirling fakery, signaling the reader to reexamine the premises of the outlined education.

For Plato really did not mean this education to be taken literally; Socrates is even teasing his own students, leading them into contradictions so they might see the absurdities of their positions. After all, their eyes and ears are also directed solely to what Socrates permits them to see and hear; their questions show their submersion in a world of illusion. They are bound, watching images and shadows cast by the clever philosopher. The "noble lie" is, in this weird way, true: the education is taking place "underground," in a cave. In fact most of the quotations Socrates says should be excised from the poets to protect his ideal citizens from corruption are actually descriptions of the cave world; Socrates explicitly returns to one of those quotes when describing the cave later. Plato was challenging the reader to follow Socrates out of the cave, as his interlocutors in the dialogue do not.

The real education must happen later, once the citizens realize that they have been following shadows. That is the task of philosophical education, which Plato has Socrates elucidate with a geometric image of the cave: the famous metaphor of the divided line.

Imagine a line cut into two unequal segments: one segment represents the visible objects and the other the objects of the realm of thought. Each segment is then cut in a similar proportion. The visible realm is divided into a realm of images and shadows—the lowest level—and a realm that contains "that of which this first is the likeness," as Socrates puts it—the solid objects of the visible world. These images and objects are found in the cave world, which is, Socrates suggests, the world in which we live. The upper portion of the line is more complicated, for it can represent the world outside the cave, lifted to a metaphysical level: the intelligible realm is the world lit by the sun of reason; it allows us to make sense of all other worlds:

	ELEMENTS OF EACH WORLD	WAYS WE COME TO KNOW THESE ELEMENTS	
	forms	intellection	d
THE INTELLIGIBLE (I)			
	mathematical objects	thought	c
	things	trust	b
THE VISIBLE (V)			
	images	imagination	a

Socrates is quite clear about the proportions of the line and the parallel relations he establishes: the ratio of the image world to the object world is the same as the ratio of the mathematical world to the form world; this ratio also appears between the visible realm and the intelligible realm ($a/b = c/d = V/I$). These ratios are actually analogies. But Socrates is interested less in the individual elements of the object world or the mathematical world than in how we *move* from one to another. For example, the ways we move from the visible to the invisible, from a realm containing objects in the world to a realm containing abstractions about the world, is similar to the ways we move from the image world to the material world, from images of objects to things themselves. What does this mean?

We know from the cave image some of what is involved in moving from shadows to objects; we know that the objects we begin with (the shadows) are dutifully studied until, suddenly, through a liberating turn of the head, the sources (the puppets) are seen. At first it is unclear which are objects and which shadows. Then it becomes obvious: the puppets are the source for the shadows; the shadows are projections of the puppets, special cases of their fuller existence. The puppets are no longer considered objects but images of something higher.

This is also true when we pass from the visible to the intelligible realm. Socrates explains, for example, that in the mathematical world the soul uses "as images the things that were previously imitated." That is, in this region, the soul examines the objects of the visible world and instead of treating them as objects and things, begins to see them as *images*, representations of something else. This is the typical pattern with which Socrates moves from one section of the line to the next. Though each part of the line seems to derive its existence from the one above (shadows could not exist without objects), the movement of knowledge is upward from images to objects, with the objects then becoming images for their successors. The process keeps repeating as the upward movement proceeds: each part of the line uses as images the things that were taken as objects in the realm below. The objects of a lower world are always examples, special cases of more general principles from a higher world. Each part of the line actively abstracts from objects below to find new objects with greater generality.

When this process is applied to ordinary objects, the first step is to determine essential qualities of these objects (such as shape or weight); these abstract qualities then become the subject of a more profound investigation to discover the general laws governing their relationships—the way we explored surfaces and spaces in the second chapter. This exploration creates mathematical objects out of ordinary ones. In the mathematical world—the world of "geometry and its kindred arts"—models are created that define abstract aspects of ordinary objects and their relations. This is also the kind of exploration that takes place within music: objects of the sonic universe (objects that are, on their own, mere "things") become subject to examination and abstraction, creating a new intelligible object, the composition. The systematic life of music and mathematics is contained in this first section of the intelligible realm.

But the highest part of the line, the Form world—corresponding to the world of the heavens and the sun in the cave analogy—is the most important and the most mysterious. It is also our unreachable goal. It is usually called the realm of Platonic Forms. The real nature of the Forms, though, is perplexing. As we have seen, we can come to understand Forms only by analogy and metaphor, only by following the proportions established by the line. Usually they are considered ideal objects of a kind, the Form of the Triangle being the source for all examples of triangles; but this interpretation, creating a static world of so-called Forms, is inadequate. Only the visible realm, V, is as static as the Forms are usually pictured to be; it is composed of sensible elements, images and things that are apprehended through simple vision. But in the higher intelligible realm, *I*, the objects are actually not sensible elements but intellectual processes. They are known not through sight but through thought; they are known through arguments, speech, or, as Socrates puts it, "logos."

The objects of the mathematical world, for example, are known through processes of reasoning, hypothesis, and abstraction much like the ones we have been so concerned with during this journey. The world of Forms goes one step further. Systems of mathematical reasoning are treated as special cases of something else happening above them, reflections of some aspect of thought of which the reasoning processes of the mathematical world are but shadowy projections. In the Form world, Socrates cryptically says, "argument itself grasps with the power of dialectic."

So we move from geometry and related arts (including, of course, music) to knowledge by treating mathematical and musical reasoning as images of something else. We take these processes and treat them as examples, special cases, of other forms of argument. We look for analogies and images. At first, it is unclear if we are moving upward or downward; if we are finding images or objects. But then differences become apparent. Arguments in the Form world do not move by "hypothesis" as do the arguments in the mathematical world; they do not proceed step by step, based upon assumptions. They proceed according to "dialectic."

Dialectic, judging from Plato's line, requires learning first by recognizing patterns, imitating them, finding their laws and then developing a model for them; it then takes the models as real objects, treating the original objects as images. Ordinary argument by "hypothesis" is just an image of dialectic, a particular example of a broader form of inquiry based on metaphor and analogy. We have found this process in use within mathematics and music: it is how we come to understand a composition, examining its premises and laws, finding their replications and transformations, learning to abstract further to see the relationship between larger patterns and more elaborate models. It is how we hear music as beautiful and how we feel it to be "true." In mathematics, the syllogistic reasoning of the mathematical world is useful only up to a point; to move forward we must step back, see what premises and arguments are being made, find similarities and so reach ever-increasing levels of abstraction. We have also used this procedure throughout our examination of music and mathematics and throughout our attempt to find Forms of which music and mathematics are merely reflections. Dialectic is the process that Plato advocates; it is a process we have tried to imitate. It requires metaphor, image, abstraction, and comparison.

Once completing the movement upward, Socrates notes, dialectic also involves a return, downward through the line. The descending philosopher no longer becomes absorbed by the images of images of the lower worlds; instead he sees them as distant representations and interpretations of primary Forms. Looking again at the lower worlds is like being presented with a completed theory and then finding its applications; it is like being presented with a composition and then finding its varied interpretations easily accessible, special cases of an abstract whole. The descent in Plato's cave journey is, we might argue, the mapping of abstraction into the concrete. Music is given meanings, mathematics applications. The Forms are made manifest in daily life. The mapping of the world is complete.

Mathematicians and musicians may spend most of their time in the mathematical world of hypothesis and reason, but the inner life of their arts is in the world of the Forms, in the processes of dialectic and its argument by metaphor. That is where our journey arrives as well. But it has no end; with metaphors we never reach a conclusion; the world is always expanding, always contracting, its extravagant wealth shaped into exquisite order. We are sounding strings whose resonances echo up and down the line, in all our caverns, as we all seek to carry knowledge further, in ever-higher argument, knowing that the end of one journey is just the beginning of another.

. . .

THE POET, we recall, was intent on seeing dawn from the mountaintop. But the illumination offered him is far more profound. The poet's head is bent downward as he ascends but then, suddenly, he is halted in his tracks not by the dawn, which he expected, but by the sight of the moon hanging naked in a clear sky, its light illuminating the "ethereal vault." At his feet he sees a "silent sea of hoary mist," while far off, "the solid vapors stretched, / In headlands, tongues, and promontory shapes," toward the distant ocean. The mist, the moon, the sky, and the ocean are each distinct objects, each seemingly subject to its own law, possessing its own character. But they are also tied together, exercising powers and influences on one another. Even the silent sea of mist is usurped by another element in the astonishing scene, for out of "A fixed, abysmal, gloomy, breathing-place," "Mounted the roar of waters, torrents, streams / Innumerable, roaring with one voice!"

The poet is stunned by the picture before him: the moon above, the mists and sea below, the relationships between all the elements, and the way the entire scene seems to evoke something beyond simple apprehension. Upon reflection, the entire scene becomes a metaphor:

> There I beheld the emblem of a mind
> That feeds upon infinity, that broods
> Over the dark abyss, intent to hear
> Its voices issuing forth to silent light
> In one continuous stream.

The emblem of the mind seems at first to be the moon itself, which does indeed seem to be brooding over that breathing place, entering into a complicated relationship with it. Like a mind, beginning the act of creation, it gives out a silent light, illuminating the vapors; then as if in response, the voices respond, issue forth, speaking with a single voice. The poet has found relationships between the objects before him; they are transformed, revealing new meanings.

But there is an ambiguity in the poem. The poet's imagination also treats the scene, with its collections of objects and relations, as a single object with a tumultuous interior life. The moon, the vapors, the voices are all contained within a larger whole. It is as if the poet were lifting the scene up Plato's line, all previous objects becoming images of something else. The entire scene—not just the moon—is also an emblem of a mind, a "type/Of a majestic intellect." It reveals its inner life, the interaction between its parts.

The scene may even be the creation of an even higher Mind: Nature itself.

> One function, above all, of such a mind
> Had Nature shadowed there, by putting forth,
> 'Mid circumstances awful and sublime,
> That mutual domination which she loves
> To exert upon the face of outward things.

As the poet watches, he, too, finds himself part of the "mutual domination." He is moved by what he sees and by the image of the Mind it provokes. The observer becomes part of a chain of worlds, a sort of Platonic/poetic divided line, in which Nature and the Scene and the Poet and his Mind are intertwined; the poet feels something taking place within the scene and within himself as he watches.

> The power, which all
> Acknowledge when thus moved, which Nature thus
> To bodily sense exhibits, is the express
> Resemblance of that glorious faculty
> That higher minds bear with them as their own.

The power impressed upon him by nature resembles the creative faculty he himself possesses. There is a chain of creative forces between these realms, a series of influences, each creating "kindred mutations" within the other. Nature bodies forth the scene and the scene exerts power on the mind; so too does the poet create kindred works. He is imitating Nature at work, reproducing in his creations the emblems that Nature had bodied forth in hers.

Such "higher minds" as his, who devote themselves to the act of creation, are no more prepared for what emerges from their labors than he was prepared for the sublime scene lying before him at the top of Mt. Snowdon. The goal in creating kindred mutations is simply to continue the analogies, to create new emblems of the mind. A mathematician will spin out a new theory or a composer create a miniature sonic universe; a poet will turn an experience into metaphor, a scene into a source of illumination. And each creator will, "mid circumstances awful and sublime," be as astonished by the result as was Kepler or Bach. Such "higher minds" as his

> *for themselves create*
> *A like existence, and, whene'er it dawns*
> *Created for them, catch it, or are caught*
> *By its inevitable mastery.*

The creators are caught by their own emanations, like the moon illuminated the roaring waters, as if the creations were discoveries rather than inventions. They are deeply moved. They hover above their creations

> *Like angels stopped upon the wing by sound*
> *Of harmony from Heaven's remotest spheres.*

Ghyka wrote that "the master minds in our Western Civilization have been, since Plato, the ones who have perceived the analogies, the permanent similarities, between things, structures, images." This is not true only of creators; it is also true of climbers, observers, critics, readers. All who approach the works reexperience those forces and sounds and find the condensed analogies within them resonating in the mind. In learning, we repeat the Creation. In listening, we hear the "roar of water, torrents, streams / Innumerable, roaring with one voice!" And in creating we hear the mind's own "voices issuing forth to silent light / In one continuous stream."

The life of the mind expands outward to the world of nature, and inward to the act of creation. It encompasses a linked set of worlds in which metaphors are resonantly connected. Mathematics and music are products of that life, "kindred mutations" that are emblems of the mind that is always creating them. They too have an inner and outer life; they too rely on metaphors and analogies, proportions and mappings. They too find a purpose in beauty and a challenge in truth. And they too resonate with other worlds up and down the divided line, from the "dark abyss" to "Heaven's remotest spheres." They remain mysteries, seeming too close to Truth to be merely human, too close to invention to be divine. They are, however,

> So molded, joined, abstracted, so endowed
> With interchangeable supremacy,
> That men, least sensitive, see, hear, perceive,
> And cannot choose but feel.

Q.E.D.
S.D.G.

SELECTED BIBLIOGRAPHY

MUSIC

Dahlhaus, Carl. *Esthetics of Music*, trans. William Austin. Cambridge: Cambridge University Press, 1982.

————. *Foundations of Music History*, trans. J. B. Robinson. Cambridge: Cambridge University Press, 1983.

Epstein, David. *Beyond Orpheus: Studies in Musical Structure*. Cambridge, Mass.: M.I.T. Press, 1979.

Forte, Allen, and Steven E. Gilbert. *Introduction to Schenkerian Analysis*. New York: W. W. Norton & Co., 1982.

Gaffurio, Franchino. *The Theory of Music*, ed. Claude V. Palisca and trans. with introduction and notes by Walter Kurt Kreyszig. New Haven: Yale University Press, 1993.

Hanslick, Eduard. *The Beautiful in Music*, trans. Gustav Cohen. Indianapolis: Bobbs-Merrill Company, Inc., 1957.

Helmholtz, Hermann. *On the Sensations of Tone*, trans. Alexander J. Ellis. 1885. Reprint. New York: Dover, 1954.

James, Jamie. *The Music of the Spheres*. New York: Grove Press, 1993.

Jeans, Sir James. *Science and Music*. 1937. Reprint. New York: Dover, 1968.

Kerman, Joseph. *Contemplating Music: Challenges to Musicology*. Cambridge, Mass.: Harvard University Press, 1985.

Kramer, Lawrence. *Music as Cultural Practice 1800–1900*. Berkeley: University of California Press, 1990.

Le Huray, Peter, and James Day, eds. *Music and Aesthetics in the Eighteenth and Early Nineteenth Centuries*. Abridged edition. Cambridge: Cambridge University Press, 1988.

Lewin, David. *Generalized Musical Intervals and Transformations*. New Haven: Yale University Press, 1987.

McClary, Susan. *Feminine Endings: Music, Gender, and Sexuality*. Minneapolis: University of Minnesota Press, 1991.

Meyer, Leonard B. *Emotion and Meaning in Music*. Chicago: University of Chicago Press, 1956.

————. *Explaining Music: Essays and Explanations*. Chicago: University of Chicago Press, 1973.

————. *Music, the Arts, and Ideas: Patterns and Predictions in Twentieth-Century Culture*. Chicago: University of Chicago Press, 1967.

Nettl, Bruno. *The Study of Ethnomusicology: Twenty-nine Issues and Concepts*. Urbana: University of Illinois Press, 1983.

Rameau, Jean-Philippe. *Treatise on Harmony*, trans. Philip Gossett. New York: Dover, 1971.

Rosen, Charles. *The Classical Style: Haydn, Mozart, Beethoven*. New York: W. W. Norton & Co., 1972.

————. *Sonata Forms*. New York: W. W. Norton & Co., 1980.

Salzer, Felix. *Structural Hearing: Tonal Coherence in Music*. 1952. Reprint. New York: Dover, 1982.

Schenker, Heinrich. *Harmony*, ed. and annotated Oswald Jonas and trans. Elisabeth Mann Borgese. Chicago: University of Chicago Press, 1954.

Schoenberg, Arnold. *Theory of Harmony*, trans. Roy E. Carter. Berkeley: University of California Press, 1978.

Subotnik, Rose Rosengard. *Developing Variations: Style & Ideology in Western Music*. Minneapolis: University of Minnesota Press, 1991.

Yeston, Maury, ed. *Readings in Schenker Analysis and Other Approaches*. New Haven: Yale University Press, 1977.

Zuckerkandl, Victor. *Sound and Symbol*, vol. 1, *Music and the External World*, trans. W. R. Trask. Princeton: Princeton University Press, 1956.

————. *Sound and Symbol*, vol. 2, *Man the Musician*, trans. Norbert Guterman. Princeton: Princeton University Press, 1973.

————. *The Sense of Music*. Princeton: Princeton University Press, 1959.

PHILOSOPHY

Adorno, T. W. *Aesthetic Theory*, trans. C. Lenhardt. London: Routledge & Kegan Paul, 1984.

————. *Introduction to the Sociology of Music*, trans. E. B. Ashton. New York: Seabury Press, 1976.

Benacerraf, Paul, and Hilary Putnam, eds. *Philosophy of Mathematics*. 2d ed. Cambridge: Cambridge University Press, 1983.

Bernardete, Seth. *Socrates' Second Sailing*. Chicago: University of Chicago Press, 1989.

Eagleton, Terry. *The Ideology of the Aesthetic*. Oxford: Basil Blackwell, 1990.

Ghyka, Matila. *The Geometry of Art and Life*. 2d ed. 1946. Reprint. New York: Dover, 1978.

Goodman, Nelson. *Languages of Art*. Indianapolis: Hackett Publishing Company, Inc., 1976.

Guyer, Paul. *Kant and the Claims of Taste*. Cambridge, Mass.: Harvard University Press, 1979.

Hofstadter, Albert, and Richard Kuhns, eds. *Philosophies of Art and Beauty*. New York: Random House, 1964.

Kant, Immanuel. *The Critique of Judgement*, trans. James Creed Meredith. Oxford: Oxford University Press, 1952.

————. *Observations on the Feeling of the Beautiful and Sublime*, trans. John T. Goldthwait. Berkeley: University of California Press, 1965.

Kitcher, Philip. *The Nature of Mathematical Knowledge*. Oxford: Oxford University Press, 1983.

Klein, Jacob. *A Commentary on Plato's Meno*. Chapel Hill: University of North Carolina Press, 1965.

Lakatos, Imre. *Proofs and Refutations: The Logic of Mathematical Discovery*. Cambridge: Cambridge University Press, 1976.

Langer, Susanne K. *Philosophy in a New Key*. 3d ed. Cambridge, Mass.: Harvard University Press, 1957.

————. *Feeling and Form*. New York: Charles Scribner's Sons, 1953.

Lévi-Strauss, Claude. *The Raw and the Cooked*, trans. John and Doreen Weightman. New York: Harper & Row, 1969.

Plato. *The Republic*, trans. Allan Bloom. New York: Basic Books, 1968.

Santayana, George. *The Sense of Beauty*. 1896. Reprint. New York: Dover, 1955.

Scholem, Gershom G. *On the Kabbalah and Its Symbolism*, trans. Ralph Manheim. New York: Schocken Books, 1965.

Strauss, Leo. *The City and Man*. Chicago: University of Chicago Press, 1964.

Toulmin, Stephen. *The Return to Cosmology*. Berkeley: University of California Press, 1982.

Voeglin, Eric. *Plato*. Baton Rouge: Louisiana State University Press, 1966.

Wittgenstein, Ludwig. *Tractatus Logico-Philosophicus*, trans. D. F. Pears and B. F. McGuinness. London: Routledge & Kegan Paul, 1961.

————. *Lectures on the Foundations of Mathematics, Cambridge, 1939*, ed. Cora Diamond. Ithaca: Cornell University Press, 1976.

————. *Remarks on the Foundations of Mathematics*, ed. G. H. von Wright, R. Rhees, and G.E.M. Anscombe, and trans. G.E.M. Anscombe. Reprint. Cambridge, Mass.: M.I.T. Press, 1967.

MATHEMATICS AND SCIENCE

Aaboe, Asger. *Episodes from the Early History of Mathematics*. New York: Random House, 1964.

Aspray, William, and Philip Kitcher, eds. *History and Philosophy of Modern Mathematics*. Minneapolis: University of Minnesota Press, 1988.

Barrow, John D. *The World Within the World*. Oxford: Oxford University Press, 1988.

Beckmann, Petr. *A History of Pi*. New York: St. Martin's Press, 1971.

Bell, E. T. *Men of Mathematics*. 1937. Reprint. New York: Simon & Schuster, 1986.

Boyer, Carl B. *The History of the Calculus and Its Conceptual Development*. 1949. Reprint. New York: Dover, 1959.

————. *A History of Mathematics*. 1968. Reprint. Princeton: Princeton University Press, 1985.

Campbell, Douglas M., and John C. Higgins, eds. *Mathematics: People, Problems, Results*. 3 vols. Belmont, Calif.: Wadsworth International, 1984.

Chandrasekhar, S. *Truth and Beauty: Aesthetics and Motivations in Science*. Chicago: University of Chicago Press, 1987.

Chinn, W. G., and N. E. Steenrod. *First Concepts of Topology*. Washington, D.C.: Mathematical Association of America, 1966.

Dantzig, Tobias. *Number: The Language of Science*. 4th ed. New York: The Free Press, 1954.

Davis, Philip J., and Reuben Hersh. *The Mathematical Experience*. Boston: Birkhäuser, 1981.

Dedekind, Richard. *Essays on the Theory of Numbers*, trans. Wooster Woodruff Beman. 1901. Reprint. New York: Dover, 1963.

Edwards, C. H., Jr. *The Historical Development of Calculus*. New York: Springer-Verlag, 1979.

Euclid. *The Thirteen Books of Euclid's Elements*, trans. Sir Thomas L. Heath. Vol. 1. Reprint. New York: Dover, 1956.

Hadamard, Jacques. *An Essay on the Psychology of Invention in the Mathematical Field*. 1949. Reprint. New York: Dover, 1954.

Hardy, G. H. *A Mathematician's Apology*. Cambridge: Cambridge University Press, 1940.

Hofstadter, Douglas R. *Göedel, Escher, Bach: An Eternal Golden Braid*. New York: Basic Books, 1979.

Huntley, H. E. *The Divine Proportion: A Study in Mathematical Beauty*. New York: Dover, 1970.

Kline, Morris. *Mathematics in Western Culture*. London: Oxford University Press, 1953.

————. *Mathematics and the Search for Knowledge*. London: Oxford University Press, 1985.

Kramer, Edna E. *The Nature and Growth of Modern Mathematics*. 1970. Reprint. Princeton: Princeton University Press, 1982.

Lawlor, Robert. *Sacred Geometry*. London: Thames & Hudson, 1982.

Manheim, Jerome H. *The Genesis of Point Set Topology*. New York: Macmillan, 1964.

Massey, William S. *Algebraic Topology: An Introduction*. New York: Harcourt, Brace and World, Inc., 1967.

Newman, James R., ed. *The World of Mathematics*. 4 vols. New York: Simon & Schuster, 1956.

Olds, C. D. *Continued Fractions*. Washington, D.C.: Mathematical Association of America, 1963.

Pedoe, Dan. *Geometry and the Visual Arts*. 1976. Reprint. New York: Dover, 1983.

Péter, Rózsa. "Mathematics Is Beautiful," trans. Leon Harkleroad. *The Mathematical Intelligencer*. Vol. 12, No. 1 (1990).

Rucker, Rudy. *Infinity and the Mind: The Science and Philosophy of the Infinite*. Boston: Birkhäuser, 1982.

Rudin, Walter. *Real and Complex Analysis*, 2nd ed. New York: McGraw-Hill, 1974.

Smith, D. E. *History of Mathematics*. 2 vols. 1925. Reprint. New York: Dover, 1958.

Spanier, Edwin H. *Algebraic Topology*. New York: McGraw-Hill, 1966.

Struik, Dirk J. *A Concise History of Mathematics*. 4th ed. New York: Dover, 1987.

————, ed. *A Source Book in Mathematics, 1200–1800*. 1968. Reprint. Princeton: Princeton University Press, 1986.

Weeks, Jeffrey R. *The Shape of Space*. New York: Marcel Dekker Inc., 1985.

Zippin, Leo. *Uses of Infinity*. Washington, D.C.: Mathematical Association of America, 1962.

INDEX

John McCleary

ed. Philadelphia native

PhD from TEMPLE (Phila)
Lasalle Colge, B.S.
field - TOPOLOGY   ask John
alga - history of mathematics   for dept
      rubber sheet geometry

V.C.
   - Math dept
   - since 1979
   - general undergrad coursework.

/ music -
       plays keyboards, piano organ
              guitar   (self taught
   Raymond Ave Ramblers
   music theory

/ hobbies: tennis
           caligraphy

_____

Brian Mann
   UC Berkley + U. Edinbourgh
          music
   MA, PhD Berkeley
         musicology. 16rc. Ital madrigal
       edits original manuscripts of same
   jazz pianist, self taught
@ VC since '87   teaches music
                              history